LIFE BY

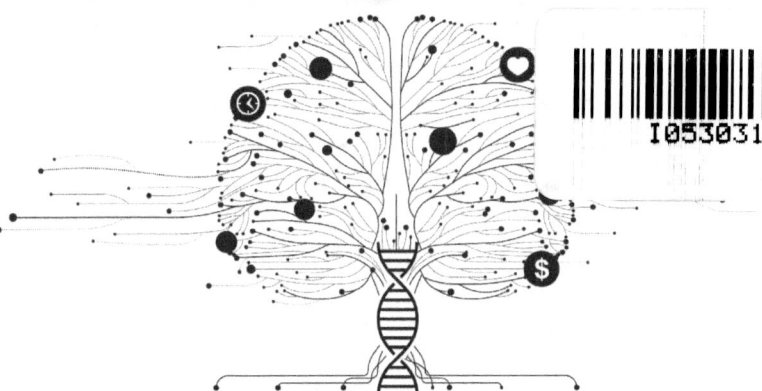

DESIGN

Automate to Master Time, Health, Money, and Unlock Personal Success

MARLON BUCHANAN

Life by Design: Automate to Master Time, Health, Money, and Unlock Personal Success

Copyright © 2025 by Marlon Buchanan.

HomeTechHacker.com

ISBN: 978-1-958648-03-2 (paperback)
ISBN: 978-1-958648-06-3 (hardcover)
ISBN: 978-1-958648-04-9 (ebook)
ISBN: 978-1-958648-05-6 (audiobook)

Edited by Graham Southorn

DEDICATION AND ACKNOWLEDGEMENTS

The challenge of finding the time to write a book that covers how to save time is an irony not lost on me. The lessons I've learned from many successful people about using automation to save time and to improve my health, finances, work productivity, and more have been invaluable. There are too many people to name individually, but I'd like to thank them all. I wouldn't be able to write books and achieve the successes I have without them.

I'd also like to thank my wife, Adriana, who supports me in this mission to write books and run a business aimed at helping others. Together we are raising a family and navigating successful careers while having many activities on the side. We must support and help each other to achieve all this, and I am extremely appreciative of her support.

I'd like to dedicate this book to everyone, strangers and acquaintances alike, who have sent me a kind word about one of my previous books, an article I've written, or a podcast I've spoken on. I do all these to help people, and knowing that I am is what keeps me going. Your encouragement means the world to me.

TABLE OF CONTENTS

OTHER BOOKS BY MARLON BUCHANAN

- *The Personal Cybersecurity Manual: How Anyone Can Protect Themselves from Fraud, Identity Theft, and Other Cybercrimes*

- *The Home Network Manual: The Complete Guide to Setting Up, Upgrading, and Securing Your Home Network*

- *The Smart Home Manual: How to Automate Your Home to Keep Your Family Entertained, Comfortable, and Safe*

- *Home Wi-Fi Tuneup: Practical Steps You Can Take to Speed Up, Stabilize, and Secure Your Home Wi-Fi*

All books are available in Kindle and paperback formats. *The Personal Cybersecurity Manual*, *The Home Network Manual*, and *The Smart Home Manual* are also available as audiobooks. You can learn more about these books at MarlonBuchanan.com.

Subscribe to HomeTechHacker.com

Be sure to check out HomeTechHacker.com for free in-depth articles about home automation, home networks, cybersecurity, and more. In addition to the articles, you'll find the following free resources:

- **HomeTechHacker Technology Advisor:** Need personalized advice for home technology purchases? This tool will ask you a few questions and recommend technology specific to your needs.

- **HomeTechHacker Wi-Fi Guide:** Following this guide will ensure you have fast, stable, and secure Wi-Fi throughout your home.

- **HomeTechHacker Resource Bundle:** This free resource bundle is filled with checklists and links to online resources to help you make the most of your home tech.

- **HomeTechHacker Technology Personality Quizzes:** Take fun personality quizzes that will entertain you, inform you, and maybe teach you something about yourself.

- **HomeTechHacker Academy:** Interested in building a smart home and don't know where to start? Are you worried about your home network security or want to protect yourself from identity theft and fraud? Enroll in HomeTechHacker Academy for online self-paced courses covering smart home setups, home network security, and ways to integrate AI tools for enhanced automation and productivity.

Also, subscribe to my newsletter at HomeTechHacker.com to get the latest updates about automation and other home technology topics.

PREFACE

In today's fast-paced world, it often feels like there's never enough time to do everything we need—or want—to do. The constant juggling of work, family, personal goals, and social commitments can leave us feeling overwhelmed, stressed, and often unfulfilled. This book was born of my own struggles to regain control over my time and improve my health, finances, productivity, and many other important aspects of my life. It is a guide for anyone who feels like their time is slipping away and who has been looking for practical strategies for getting more enjoyment out of their life.

I began this journey into time management and automation several years ago. As an IT professional, I've always been intrigued by the way technology can simplify complex tasks and improve efficiency. However, I was often guilty of letting technology take over my life, leaving me constantly distracted by emails, notifications, and the ever-growing list of things to do. I realized that while technology can be a tool for productivity, it can also be a source of stress if not used wisely.

I look upon time as life's budget. In business, I am fond of telling the executives I work with: "Don't tell me your priorities. Show me your budget and I'll tell you your priorities." Our personal lives are no different. Our true priorities are based on what we spend our time doing, not on what we say they are. We should spend more time on the things we really love and consider important, and less time on the activities we don't value.

The goal of this book is to show you how to leverage systems and technology to improve various aspects of your life, with a focus on freeing up time. You'll learn how to automate routine tasks, create systems that streamline your daily life, and reduce mental clutter in areas such as money management, relationships, work productivity, health, fitness, and more. By doing so, you can make more time for what truly matters to you—whether that's personal growth, nurturing relationships, improving your well-being, or simply enjoying life.

Through practical tips, real-world examples, and my own experiences, I'll guide you step by step on a journey to optimize your time, reduce stress, and create a life that reflects your values and priorities. It's about working smarter—not working harder—making technology your ally in every area of your life.

I hope this book provides you with the tools and inspiration to automate the tasks that drain your time, so you can focus on the things that matter most and build the life you've always wanted.

ABOUT THIS BOOK

Are you feeling overwhelmed by all the requirements and stresses of life? Are you trying to have more time for yourself, your friends, and your family? Do you feel like you are missing out on important events in your life because you don't have time for them?

It feels like every moment of the day is spoken for, doesn't it? The way we use technology is to blame. At any given time, you'll see a flood of notifications on your phone telling you what you need to do right now, what you forgot to do, and all the things you need to get ready to do. You're flooded with emails from your job, your children's schools, and all those newsletters you don't remember signing up for.

Then there are the financial institutions, utilities, insurance, government agencies, and other organizations that constantly send you texts and emails about bills, privacy notices, term changes, and offers to purchase more.

Modern life is not just extremely busy but extremely fragmented. Our ancestors were busy too, but they were able to focus more of their time on fewer things. Technology has empowered us to get more things done, but at the cost of giving us more things to do.

There are few things in this world more valuable than time. Money can buy only so much more of it, and you can run out of time at any moment. That's why it's important to maximize the time you have.

And, this book isn't just about managing time—it's about automating aspects of your life so that you can improve your life. From managing your health and fitness to maintaining your finances to strengthening relationships and home maintenance, automation can help you reclaim your time and energy and improve all these areas.

When you finish reading this book, you'll know how to do the following:

- Slow down and focus on what matters most in all aspects of your life—health, relationships, finances, and more

- Leverage technology to automate tasks you don't want to do yourself

- Create systems and habits that allow you to spend more time on what you truly love, whether that's travel, fitness, or personal development

- Put your finances, health and fitness goals, home maintenance, and more on auto-pilot so you can focus on your priorities

- Improve productivity at work and in life, allowing you to spend less time working and more time living

This book is for anyone looking to simplify their life and enhance their personal well-being and success. It's about using automation and smart systems to improve your life in every area that matters to you.

You don't need any particular background to understand or benefit from this book. All you need is a desire to have more time for the important things in your life and a willingness to embrace new techniques to get you there.

Time is the currency of life. Maximize the way you use it today, but remember to invest it for tomorrow.

HOW TO USE THIS BOOK

This book steps you through the ways of creating more time for what you want to do in your life, by automating the things you don't want or need to do. It starts with a foundation of knowledge about why we are so busy. Then, using that knowledge, the book guides you through the steps needed to get the time-saving results you want. At the same time, it shows you how to use automation to achieve success in many aspects of your life. If you read this book sequentially, you'll learn what is causing the time management crisis in your life, the ways to prioritize what you want to spend your time on, and many techniques for making that happen.

Although this book is intended to be read sequentially, it works well as a reference for those wanting to brush up on and find new ways of improving their life's time management and achieving personal success. The layout and headings make it easy to quickly find the information you may be looking for.

Be sure to take advantage of additional resources in the appendix and glossary, including inspirational quotes and additional online resources. These resources will cement and augment the learnings you gain from this book as well as inspire you on your life automation and time-saving journey.

Conventions Used in This Book

Here are the conventions used in this book, which highlight important information:

- **What I Do:** In several sections of this book, I give specific information, addressing the topics in the context of my own life. In the Health and Fitness section, for example, I discuss what I've done to automate my personal health journey to lead me to better physical and mental fitness.

- **In the Real World:** Throughout the book, you'll find useful stories and case studies about automating your life and creating time for the people and activities you value most.

- **Key Takeaways:** At the end of each chapter, I provide a short list of key information that summarizes important learnings from the chapter.

CHAPTER 1

WHY ARE WE SO BUSY?

We are not that busy; we are just distracted.
—Shawn Wells

The demands for our attention are unprecedented in human history. Technology and societal changes have increased these demands exponentially in just the last two decades. Think about how much time you spend using mobile devices, computers, TVs, and other screens. You're certainly not alone because the average person spends seven hours a day on a screen[1]!

As recently as the nineties, Americans spent fewer than three hours a day on a screen. Back then, almost all our screen time was spent watching our favorite shows and news on television. Now, in the average person's seven hours, they are still watching plenty of TV but also using social media, streaming videos, and playing games. And while it's true that watching traditional TV has declined slightly, social media, streaming, and video games are more than making up for it. As a result, screen time is increasing.

Today, you are part of the attention economy. Everything on all these screens is designed to compete for your attention. Social media gives you alerts for new posts, likes, and followers. Its algorithms are designed to push content that

---

1 "17 years of your adult life may be spent online. These expert tips may help curb your screen time," Fortune, Accessed May 20, 2025, https://fortune.com/well/article/screen-time-over-lifespan/.

will keep you on that platform. This isn't necessarily the content you want to see. It is the content you'll watch and respond to, which in many cases is content that provokes a negative reaction.

Social media isn't alone in competing for your attention. Financial apps send alerts for transactions and payments. Stores send emails and notifications for the latest sales and promo codes. Schools send parents an abundance of text messages and emails telling what your child needs to study and bring to school and alerting you to upcoming activities or their being tardy. These are just a few examples of how many things are competing for your attention.

Many studies have documented what's happened to our attention span over the years. Dr. Gloria Mark, Professor of Informatics at the University of California, Irvine, and author of the book *Attention Span: A Groundbreaking Way to Restore Balance, Happiness, and Productivity,* has conducted numerous studies. In her research, she and her team tracked how often people shift their focus to something new when using electronic devices over several years. In the early 2000s, it was once every two and a half minutes. Now, it's every 47 seconds. In the next few years, this will probably drop to under 30 seconds.

One explanation is that all of the notifications and constant competition for your attention teach your brain to switch contexts and crave something different more often. This training affects you even when you aren't getting the constant stimuli of being asked for your attention. You'll want to switch what you are focusing on quicker, just because your brain is accustomed to doing so.

All these demands on our attention are indicative of how many things we are expected to keep track of as modern humans. In addition, all that context-switching forces us to spend more time to get things done. According to the American Psychological Association (APA), doing more than one task at a time, or multitasking, takes a significant toll on our productivity[2]. Numerous studies have concluded that time is lost every time we switch our attention, and over the long haul that time really adds up.

The combination of having more to do, with more competition for our attention and with devices that train our brain to be less efficient at completing tasks, contributes significantly to the reason we are so busy.

Status Symbol

While technology is largely responsible for the attention economy's creating busyness for us, our general culture of working to appear better to others also drives a large part of our busyness. Many of us fall victim to the "rise and grind" culture. We are expected to work hard, make sure we are always hustling to make that extra buck, reach that extra goal, and achieve more to have a better life in the future.

#Riseandgrind is so much more than a popular social media hashtag. It is a status symbol that creates a type of peer pressure. If you are in the "rise and grind" culture, everyone is talking about how hard they are working to achieve more. The gig economy, where people can find on-demand work, has exacerbated the busyness caused by this culture. Sure, you can have your regular nine-to-five job, but rise and grind asks, "What are you doing on the side?" You are expected to be an Uber driver or make Doordash deliveries or consistently put out content on your blog, YouTube, and other media channels.

---

2 "Multitasking: Switching costs," American Psychological Association, Accessed May 20, 2025, https://www.apa.org/topics/research/multitasking.

But rise and grind is about more than making money on the side. It's also about doing more to be a better parent, a better friend, and a better and more productive employee. It is all about doing more to improve your future life. It's a badge of honor that many wear to show they are hustling to get more done all the time.

On the face of it, it is admirable. But sometimes, the pursuit of status leads people to do more and more, and they get busier and busier. They often take it too far, ending up too busy and sacrificing too much of today for an uncertain tomorrow. Unfortunately, too much rise and grind will eventually grind you down.

FOMO—Fear of Missing Out

Your friends are taking a wonderful vacation in Hawaii. Your next-door neighbors are talking about the educational after-school programs they have enrolled their children in. At the same time, you get a notification on your phone about some of your favorite social media stars going live? What do you do?

You might try to make sure your kids are also enrolled in good afterschool educational programs, maybe even the same programs your neighbors have been mentioning. You might decide it is time to book an exotic vacation. And naturally, you check your notifications constantly so you can join the live stream early.

What causes you to do all this? It's fear of missing out, or FOMO. You don't want to miss out on opportunities that others are taking advantage of. FOMO is deeply ingrained in the human psyche. We have a horrible feeling that we are missing out because of the special pressures we feel to be a part of something. It is a form of peer pressure.

The anxiety produced by FOMO causes us to spend time making sure we aren't missing out. We work hard to make sure we won't miss out on the experiences and opportunities others are talking about. Sometimes, we don't stop to think hard enough about whether we need to be a part of these activities or whether they are the best use of our time. Unfortunately, our desire to keep up with others and not miss out on experiences causes us to add even more tasks to our daily lives.

The Paradox of the Abundance of Choices

We have so many choices to make. We also have a lot of options when making a choice. On the surface, this seems like a good thing, and it certainly can be. After all, wouldn't you rather have three good options when buying a new car instead of just one? We have been taught to work hard to increase the choices and opportunities in our lives.

As a person who spent a lot of time around friends and acquaintances that had more choices than me, I understand the drive to create more opportunities. In fact, this lack of choices was what drove me to a lot of the successes I've had in my life. I knew I had to work hard to create opportunities to be successful. This type of choice creation is mostly good.

The problem comes with all the day-to-day choices we have to make. What's for dinner? Which colleges should my child apply to? Which brand of toothpaste is best? What show should we watch on TV tonight? Which dryer should we buy to replace the one that broke in our house? What contractor should we hire to remodel our kitchen?

All these choices pile up and cost us time. We have to research which option is best, and often spend time discussing and debating the right options with our friends and family.

Have you ever come home from a hard day at work and just wanted everything decided for you? You just want someone else to pick dinner or to pick the evening's activity. This is because constantly picking from an abundance of choices is time-consuming and exhausting.

I'm not saying having choices is a bad thing. The problem is that we have too many choices for too many different things. We spend too much time evaluating choices, even when many of our choices will satisfy our needs. How much time have you spent reading reviews and running Google searches for a simple product you are going to buy on Amazon? Why did you do that? Because Amazon had so many choices, and you wanted to be sure you bought the right one.

We often think that having many choices will help us make better decisions. In reality, often we'd be better off with having just one or two good choices that fit our needs.

We Are Busy for the Sake of Being Busy

Another reason we are so busy is that we are simply accustomed to being busy. It's a habit we've built up over time, especially in the United States. We tend to fill our free time with activities, often because we feel unproductive if we aren't always doing something. It's akin to the "rise and grind" culture I referenced earlier.

One of the problems is that we occupy our free time with "fillers." These fillers are often activities that don't lead to significant accomplishments. They just help us pass the time. The problem is that we need rest in order to work towards higher priority achievements. The fillers often lead to our being more stressed and having less mental capacity when we finally do have time to work on our priorities.

Why do we do this? A lot of our self-worth is tied up in being busy. Being busy means you get to tell other people that you are busy and that you are in demand. No one wants to be looked at as someone sitting around doing nothing. Also, to some degree, we have a fear of being judged for what we haven't done. Being busy is the ultimate excuse for not getting other things done. Tim Ferris, the author of the uber-popular book *The 4-Hour Workweek*, summed it up best with the following quote: "Being busy is a form of laziness—lazy thinking and indiscriminate action. Being busy is most often used as a guise for avoiding critically important but uncomfortable actions."

Over the years, technology has made us much more efficient. According to the Economic Policy Institute, our work productivity rose 61.8% from 1979 to 2020[3]. Although this statistic doesn't cover our personal lives, it is an indicator that, overall, we can get a lot more done in less time. Unfortunately, we've used the time saved through productivity gains to do even more, and technology makes it easier for us to do that.

Information and activities are available 24-7. If you're not careful, you could find yourself in a never-ending state akin to being trapped in a Las Vegas casino. There's no need to leave the casino in order to do or access anything you want. Sometimes, you don't know how long you've been there or even where the exit is.

The key to breaking out of this pattern is to realize that activity does not necessarily equal achievement. You can accomplish more if you take time to not be busy. We'll discuss exactly how throughout this book.

---

3 "The Productivity–Pay Gap," Economic Policy Institute, Accessed May 20, 2025, https://www.epi.org/productivity-pay-gap/.

Key Takeaways

- **Attention economy:** Modern technology, including social media, apps, and constant notifications, competes for our attention, making it harder to focus on what's important.

- **Multitasking myth:** Studies show that multitasking reduces productivity and increases time lost due to context switching.

- **Rise and grind culture:** The pressure to constantly work harder and be busy has become a social norm, often leading to burnout and less time for meaningful activities.

- **FOMO (Fear of Missing Out):** The anxiety about missing out on opportunities or experiences, fueled by social media, leads us to fill our lives with unnecessary tasks.

- **Choice overload:** The sheer volume of choices we face daily, from what to eat to what products to buy, drains mental energy and time.

CHAPTER 2

WHAT DO YOU *REALLY* WANT TO BE BUSY DOING?

The key is not to prioritize what's on your schedule,
but to schedule your priorities.
—Stephen Covey

I'm going to let you in on a secret: I think it's okay to be busy. I know what you're thinking: I've just spent a chapter explaining why we're all so busy, and none of my explanations implied that being busy is good. Most of them focused on negative external influences on our lives, which cause us to be busy.

But the real issue isn't that we are busy or that being busy is necessarily bad. The real issue is that being busy eats up time. Time is one of our most precious gifts and it is finite. You can make more money, more friends, and more memories, but you have to spend time to do any of these. You can spend time, but you can't keep it. And, once you've spent or lost time, you can never get it back.

Are there things you can do to give yourself more time? Maybe. You can be healthier and live safer. But you can never reclaim time you've already spent, and you are not promised any set amount of time no matter how healthy or safe you are.

Because of how precious time is, life, in large part, should be about optimizing how you spend it. Ultimately, you should aim to feel good about what you've spent your time doing. Optimizing time isn't just about being as productive as you can be at any given time. It's about spending time on what you love and what gives you fulfillment.

Scenario 1: What if you had $500,000 and knew that you *wouldn't* be able to make any more money once you spent that amount?

Scenario 2: What if you had $500,000 and knew that you *would* be able to make more money once you spent that amount?

Question: Would you spend the same $500,000 differently?

Answer: Of course! In the first scenario, you'd try to spend $500,000 on things that are really important to you and not to waste any of it on things you don't really care about.

When it comes to spending time, many people approaching retirement apply this lesson to the time left in their lives. Many are in positions where their financial resources are fixed, and they've lived long enough to know that their time left is limited and that they can't get back time they've already spent.

For many years, I enjoyed watching the TV show *Arrow*. One of the main cast members was Katie Cassidy, the daughter of David Cassidy, best known for his role as Keith Partridge on the hit 1970s TV show *The Partridge Family*. When David passed away, Katie shared with her fans her father's last words: "So much wasted time." Those words have stuck with me. In the end, all we have in life comes directly from how we spend our time. As Katie said when reflecting on her father's last words, we should "never waste another minute[4]."

To fully embrace a life by design, the first step is to determine what you truly want to spend your time on. We all have tasks that drain our energy, take up precious hours, or simply don't add enough value to warrant our attention. It's essential to take a step back and evaluate your life—what matters most to you? What activities bring you joy, fulfillment, or progress toward your long-term goals? What tasks do you wish you could eliminate or minimize from your daily routine?

Once you've identified these priorities, the next step is automating everything else. This is where the true power of automation lies: It's not just about making life easier; it's about freeing up time for the things that align with your values and passions. By automating tasks you don't enjoy or those that take up too much of your mental space—whether it's managing household chores, finances, or even some aspects of your health—you can create more room for the things that truly matter.

What Should You Spend Your Time On?

So if you accept that time is precious and not to be wasted, how do you decide what to spend your time on? This question asks you to figure out what is important to you. There are more things to do in life than anyone has time for. Trying to get so many things done is one of the reasons we all feel so busy. But what should you be busy doing?

One way to approach this is to better understand the nature of the activities on which you currently spend time. Make a

---

4 *"David Cassidy's Last Words Are Shared by His Daughter,"* Time, Accessed May 20, 2025, https://time.com/5036891/david-cassidy-last-words-katie-cassidy/.

list of the regular activities you spend time on from month to month. At this point, don't worry about how much time you spend on them, just so long as each activity is something that, you think, takes up a noticeable amount of time.

Then you'll want to put these things in a couple of categories:

- Activities you *have* to spend time on
- Activities you *want* to spend time on

You define what you *have* to spend time on. For example, do you *have* to spend time on yard work? Of course—if no one else in your house can do it and you can't afford regular yard maintenance. But if someone else can do it, it's not something you *have* to do. On the other hand, going to work is probably something you *have* to do, unless you don't need the money.

Now consider things you *want* to spend time on. Let's say you normally drive on your commute to work. Do you *want* to be spending your time driving there and back? Your answer might be yes. But what if there are other ways of getting there, such as public transportation, or you have the option to work from home? Then you don't *have* to spend time driving. And if you don't enjoy driving or taking public transportation, you neither *have* to nor *want* to spend any time behind the wheel or taking public transportation.

An easier way to represent this is by creating a chart of monthly activities. Although I recommend that you create this list of tasks in a program like Google Keep or Notion (more on those tools later), I've provided a more visually attractive representation (Fig. 1).

MONTHLY TASKS

H = Have to do
W = Want to do
N = Neither

Task	H	W	N
Working	X		
Driving to work			X
Spending quality time with spouse		X	
Playing with children		X	
Attending child activities		X	
Maintaining the lawn	X		
Managing family investments		X	
Cleaning the house			X
Doing laundry	X		
Managing home climate			X
Grocery shopping			X
Paying bills			X
Volunteering		X	
Maintaining friendships		X	
Exercising		X	

Figure 1. Rate monthly activities as essential, enjoyable, or neither.

Figure 1 provides examples of different monthly tasks and the way I've categorized them. A complete diagram would be much longer.

Now use the information you've gathered from this activity to identify the items you don't *want* to do and don't *have* to do. These are the items you should be working to eliminate from your monthly activities. If you can't eliminate them, you should find ways to spend less time on them. We'll discuss techniques for automating and outsourcing later in this book.

So far, we've looked at your life today. But what would you prefer your life to be like in future? And crucially, how should you change the tasks you do today to accomplish this transformation?

Look to the Future

You've heard of a "bucket list." It's the list of things you want to do before you die, or "kick the bucket" as it's sometimes called. Bucket lists tend to be filled with large and ambitious goals and experiences. The younger you are, the farther down the road they are.

I like the idea of a bucket list—it's a fun way of imagining all the good times you're going to have. But the biggest problem is that there is no plan or timeline associated with the ways you are going to achieve the items on the list.

Bucket lists would be much more useful if the items were put into buckets by age range. For example, instead of saying that you want to travel to France sometime in the future, you state that you want to travel to France between the ages of 36 and 40. Instead of saying that you want to one day buy a house, say that you want to buy your own home between the ages of 31 and 35.

Another problem with bucket lists is that they focus on your long-term dreams. But often there are practical and shorter-term goals you want to achieve, like losing weight or gaining muscle. You should put these types of items into age-based buckets too.

To keep things manageable, make the buckets span five years each. So, if you are 36 years old, you'd have buckets for 36–40, 41–45, 46–50, and so on. Think about what you want your life to look like during each of those age ranges. Your buckets should describe the things you want to do and accomplish in order to have that life. Make sure the buckets comprise only items you deem important.

I recommend using an online note-keeping or task management tool, but you could create a more visually attractive template instead, like the one shown in Fig. 2. Depending on your age, you may have more buckets and/or different age ranges for your buckets.

BUCKET LISTS

26–30 Bucket
Buy a Home
Take Vacation in Australia
Learn Mandarin
Get married
Attend Coachella
Pay off student loans

31–35 Bucket
Have children
Run a marathon
Go on a solo trip to a new place
Learn photography
Go skydiving
Start my own business

36–40 Bucket
Take a big family vacation
Learn to play a new instrument
Have a $1M dollars invested
Attend a TED Talk
Complete a triathlon
Publish a book

41–45 Bucket
Take a sabbatical from work
Visit all seven continents
Host a large family gathering
Become an executive
Join a local board of trustees
Teach a class or workshop

Figure 2. Compile your short-term and long-term goals into age buckets.

Once you've made these buckets, it's time to make plans for the ways you will achieve them. If there are things you should be doing now to achieve the items in a bucket 20 years from now, then you have to figure out where those predecessor activities should be prioritized with all other things you already have to do today.

Then, you'll want to add these activities to the monthly activities list you created earlier. Even if all the predecessor activities aren't monthly, it's still a good idea to add them to the list. You can denote which months you will work on them and which ones you won't, and finally remove them from the monthly activities when they are complete.

What to Spend Your Time on Right Now

Up to this point, we've cataloged what you spend your time on right now. We've also worked through your future plans and the ways they affect how you should spend your time. But what about things that aren't so far in the future? Let's call them your current goals.

Current goals are things that you'd like to do but are not doing right now. For example, would you like to read more books? Spend more time with your best friends? Finish that DIY home remodel that's been lingering? The only way you can get to these items is to make them a priority in the tasks you do every month now.

Take a look at your latest updated monthly task list. As you review the list, ask yourself, "What's missing? What are the things I want to do now that aren't on this list?" These could be things like reading a book every month or having a regular dinner with close friends. It could also be something you don't enjoy but still want to accomplish, such as making progress cleaning out your garage.

Add the items you want to accomplish now to your monthly task list.

Putting It All Together

Now you have an entire list of things you *want* to do and *have* to do month to month. If you've followed the steps earlier in this book, this includes the following:

- Tasks you *have* to do each month
- Tasks you *want* to do to achieve future goals
- Tasks you *want* to do to achieve current goals

You shouldn't expect this list to be filled with only those things that you enjoy and that make you money. Some things you just *have* to do. You have to keep track and manage your finances. You need to take care of your home (cleaning, yard work, etc.). And you may have to commute to work.

The rest of this book is focused on reducing the time you spend on those things you *have* to do but don't *want* to do as well as on achieving personal success through automation. This will free up time to spend on the things you love.

Key Takeaways

- **Time is finite:** Time is one of the most precious resources. Once it's spent, it can't be reclaimed, so it should be used wisely.
- **Prioritize activities:** Focus on activities that align with your values and long-term goals. Don't simply do more.

- **Task categorization:** Split your activities into categories—(1) things you have to do and (2) things you *want* to do—and eliminate or work on automating the ones you don't enjoy or need.

- **Bucket lists:** Create actionable plans for your long-term and short-term goals by organizing them into age-based categories, so you stay focused on meaningful achievements.

CHAPTER 3

FORMULATING AUTOMATABLE HABITS

People do not decide their futures.
They decide their habits, and their habits
decide their futures.
—James Clear

Software engineers like me make a living turning what people want done into simplified tasks that can be automated in code. This automation is the key to software making things simpler and faster.

The concepts a software engineer uses to save their customers time are the same concepts you can use to simplify your life. You need to break down the things you do into simple, automatable steps. You also need to build habits that can be automated.

One way to simplify your life is to schedule many of the tasks you complete each day. If you take medicines or vitamins daily, you likely take them at the same time, or multiple times, every day. By doing this, you build a simple habit that always gets done. In fact, many of the important things you have to do each day are probably done on a schedule.

Think about arriving at work. You probably get to work at around the same time each day. And there's a whole set of scheduled activities (waking up, eating, getting dressed, commuting, etc.) that you have to do to make it happen.

As you are scheduling these activities, pay close attention to the priorities you developed in the previous section. You'll want to make sure that you schedule your most important tasks in a way they'll be most likely to get done. This doesn't mean you have to schedule all the important activities first. It just means that they should be on the schedule and that they are scheduled at times when they are likely to get done. The timing of important activities will be different for everyone. You will need to adjust your schedule as you learn the best times for getting things done.

Think about ways you can break these tasks down into simpler activities. Most likely, some of these tasks can be automated or done by someone else. Later in this book, we'll go over specific ways to automate your finances, activities around your home, and other parts of your life.

An important part of this process is writing things down. Don't just have a schedule of your activities in your head; write down your schedule of activities day by day. Having a schedule that is for each week or for an entire month ahead will suffice. Also, make sample task lists that go with each day. Your goal for each day should be to complete most or all of the tasks on your list.

As you complete your task list, give some thought as to which tasks you'd like to automate. These should be repeatable tasks that you don't want to do. They also may be tasks where the quality of the finished task, or the way it gets done, isn't very important. Below are some examples of tasks that may fit these criteria:

- Grocery shopping
- Household chores
- Paying bills
- Keeping track of subscription services
- Email management
- Fitness tracking
- Scheduling routine medical appointments
- Laundry
- Yard work
- Home maintenance
- Data backups
- Budgeting and expense tracking

It's often a good idea to focus on automating lower-priority tasks first. Over time, when you get better at automating, you'll have the skills to automate your more important tasks. Also, one of the beauties of automation is that you can gradually improve the way you automate things to perform your tasks better, faster, or both!

Leveraging Technology to Automate Your Life

Throughout human history, technology has been the driving force behind increased productivity. From the earliest wood and stone tools to today's artificial intelligence, innovations have enabled us to do more in less time. But in the last 50 years, the rate of technological improvements and adoptions has grown exponentially. In that time, we've seen the introduction of the PC, the World Wide Web, Wi-Fi, and smartphones.

Companies have entire functions dedicated to finding ways of leveraging technology to improve productivity, largely by creating automation. Car companies automate their car assembly lines with computers and machines. Online stores like Amazon automate their payment and fulfillment processes. Digital marketing professionals use technology to automate their newsletters, social media posts, and other promotional materials.

Today, automation has arrived for everyday use. Technology makes it easier for us all to get things done more quickly, and with less effort.

Basic Examples of Automating Tasks

I've written a lot about automating tasks, and the rest of this book focuses on different methods and areas of automation. But first let's take a look at a couple of examples of how simple activities can be broken down into tasks that can be then automated.

Climate Control

One of the easiest tasks to automate in your house is controlling the climate. If you have a manually controlled thermostat, you currently have to do some or all of the following:

- Set the mode to heat or A/C
- Set the temperature
- Readjust the temperature when it doesn't feel right

You may have to do this over and over again each day. And, if your home has multiple thermostats—perhaps one on every floor or even each room—you have even more manual steps to do.

Instead of making these adjustments multiple times a day, you may be able to save time by having a programmable thermostat. Some can set your home to certain temperatures at the same time each day. Others offer separate programs on weekends, or even daily. Another common feature is an "auto" setting, whereby the heat will turn on if the house temperature is below a certain level, or the A/C will turn on if the house temperature is above a specified level.

These programmable thermostats can save you time and money if you have a predictable schedule. Each day, your home will go to a certain temperature at set times. But what happens when your schedule changes? Or when there are multiple people with different schedules in your house? Then, you will still end up making multiple adjustments.

This is where a smart thermostat can come in handy. Smart thermostats can be programmed to adjust your temperature to a different schedule each day like some of the upper-end programmable thermostats. In addition, many of them can sense when someone is home and adjust the temperature, which is useful when you are off schedule or have multiple schedules to account for.

Smart thermostats can even take things a step further. Some, like Google's Nest thermostats, can learn your schedule over time so you never have to adjust the programming. These features can save you not only time but also money by ensuring that your heating and cooling systems are in use only when they are needed.

Controlling your home's climate in an automated fashion needn't start and end with a smart thermostat. If you have fans, you can convert them to smart fans that turn themselves on and off at appropriate times. For example, if you have a bidirectional ceiling fan, you can have it rotate clockwise in the winter to distribute warm air and counterclockwise in the summer to create a cooling breeze. The fans can be coordinated with your primary heating and cooling systems.

Automatic fan control is also useful for bathroom fans. These can be automated to turn on and off based on humidity, presence, or even use of the bathroom lights. I don't have a humidity sensor in my bathrooms. So, I have the bathroom fan automatically (1) come on when the light is turned on and (2) turn off 20 minutes after the light is turned off.

Your fireplace can contribute to climate control too. While you might not want your fireplace to ever turn on by itself, it can be helpful for it to coordinate with the rest of your heating system. For instance, you can configure a smart fireplace to turn off when the temperature reaches a certain level, or if the air conditioning comes on.

Air quality is another important part of your climate. Smart air purifiers and humidifiers can automatically make sure you're breathing the kind of air you want to breathe.

How does this save you time? Well, it can all happen automatically with automation and smart devices. You never have to touch or think about your thermostat, air purifiers, and humidifiers. And you don't have to worry about turning off your fireplace and fans. You'll just feel comfortable and breathe easy all the time.

Meal Planning

You already spend a significant amount of time on food. I mean not just eating but also deciding what you're going to eat for each meal and making multiple trips to the grocery store to purchase ingredients. So, automating meal planning and preparation can save you significant time and effort, especially if you do it consistently. You can break the activity down into the following steps:

1. **Create a weekly meal plan:** Set aside time each week to plan your meals for the upcoming week. Consider factors such as dietary preferences, nutritional balance, and convenience.

2. **Make a master recipe list:** Compile a list of your favorite recipes categorized by type (e.g., breakfast, lunch, dinner, snacks). This will serve as a reference when planning meals.

3. **Build a grocery list:** Based on your meal plan, create a shopping list of ingredients you'll need for the week. Organize the list by category (e.g., produce, dairy, pantry items) to streamline shopping.

4. **Batch cooking:** Dedicate a portion of your weekend or a free day to batch cooking. Prepare large quantities of staple ingredients such as grains, proteins, and vegetables that can be used in multiple meals throughout the week.

5. **Prep ingredients in advance:** Wash, chop, and portion out ingredients ahead of time to reduce prep work during the week. Store them in airtight containers or resealable bags in the refrigerator.

6. **Use slow cookers or instant pots:** Invest in kitchen appliances like slow cookers or instant pots to simplify meal preparation. These devices allow you to set and forget meals, saving you time and effort.

7. **Optimize leftovers:** Plan meals with leftovers in mind. Cook extra portions and repurpose them into

new dishes to minimize food waste and save time on cooking.

8. **Freeze meals for later:** Prepare freezer-friendly meals in advance and store them for busy days when you don't have time to cook. Soups, stews, and casseroles are great options for freezing.

9. **Stay flexible:** While it's best to stick to your meal plan, you can remain flexible and adapt it as needed based on schedule changes or ingredient availability.

10. **Track and evaluate:** Keep track of your meal plans, grocery lists, and cooking times to identify areas for improvement and streamline the process further over time.

Meal planning is an example of activity automation that doesn't rely on technology. Additionally, you can enjoy the benefits of healthier eating habits, reduced stress, and more free time in the long run.

In fact, you can use technology to make meal planning even easier, along with many other household chores. We'll discuss these examples later in the book, but in the next chapter we'll find out whether automation can help make you more money.

Key Takeaways

- **Daily routines:** Small, consistent actions are the building blocks of productive habits; automation can make these habits easier to maintain.

- **Simplification:** Break complex tasks into simple, repeatable steps that can be automated, making it easier to stay on track with your goals.

- **Automating lower-priority tasks:** Start by automating small, less significant tasks, such as grocery shopping or scheduling, to free up more time for high-priority activities.

- **Technology as an enabler:** Use apps and smart devices to track and manage your routines and habits, helping you stay consistent.

CHAPTER 4

AUTOMATING YOUR FINANCES

Wealth is largely the result of habit.
—*John Jacob Astor*

Managing your finances can seem a tedious, if necessary, task. But done effectively, it allows you to control your economic well-being, helping you navigate life's uncertainties and achieving your financial goals. You'll be able to make informed decisions about spending, saving, and investing. Ultimately, this will give you greater financial stability, freedom, and opportunities for growth. I believe that for most people, managing their finances effectively is more important than making a good salary.

Automation reduces the need for manual intervention, saving valuable time and effort that can be redirected toward other priorities. But it can do so much more. Consistent, disciplined saving and investing lead to wealth. Setting up automatic contributions to savings and investment accounts is an easy way to ensure you do this. Automatic bill payments go a long way toward making sure financial obligations are met on time, helping avoid late fees and penalties. Plus, streamlining processes reduces the likelihood of human error.

Another huge benefit of automated financial systems is the greater visibility they give you into your cash flow and spending patterns. You can then make more informed decisions about your finances and optimize your saving and investment strategies for long-term growth.

Let's get into some strategies for automating your finances.

Disclaimer: The information provided in this chapter is for informational purposes only and should not be construed as financial, investment, or legal advice. Always consult with a qualified financial professional before making any financial decisions. The author is not responsible for any financial outcomes resulting from the use of the information presented in this book.

Review Your Cash Flow and Financial Accounts

If you don't have a budget or closely track what you regularly pay and how you make each payment, now is the time to do it. You can accomplish this with a cash flow review. To do this, look through the last three to six months of bank and credit card statements. As you look through them, note the following:

- All money that's regularly going out. You'll want to document the purpose of the payment, what company the money is going to, how much the payments usually are, and how often the payments are made.

- All money coming in (e.g., paychecks, investment income, rental income, etc.).

- Any charges or deposits you aren't 100% familiar with. It is a good practice to thoroughly research these.

The chart shown in Fig. 3 gives you an idea of the information you should record as you review your cash flow. You could also track this in a spreadsheet or by using a purpose-built expense-tracking app. We'll discuss ways of automating your accounting later in this section.

CASH FLOW

Income	Source	Amount	Account	Frequency
Paycheck	Acme Corp	$3,000	Chase Bank	Bi-monthly
Rent	Rental Properties	$2,000	Chase Business Bank	Monthly
Interest	Savings	$150	Fidelity Brokerage	Monthly
Rideshare	Lyft	$100	Chase Bank	Weekly

Expense	Payee	Amount	Account	Frequency
Mortgage	Bank of America	$1,500	Chase Bank	Monthly
Groceries	Various Stores	$500	Visa	Monthly
Car Note	Chase Bank	$450	Chase Bank	Monthly
Electricity	ABC Utility	$100	Mastercard	Monthly
Mobile Phone	Verizon	$90	Visa	Monthly

Figure 3. Itemize your income and expenses to track your cash flow.

From this information, list all your bank, investment, credit card, and other financial accounts. Also, note accounts that are barely used, or that are at the workplace. You'll use this information in the next few steps.

Reduce the Number of Financial Accounts

You should simplify your finances before you start automating them. Reducing the number of financial accounts you manage can bring several important benefits. Fewer accounts mean less complexity in tracking account balances, transactions, and associated fees. This consolidation can lead to a clearer picture of your financial health, making it easier to manage your budget and savings goals. With fewer accounts to monitor, the effort required to manage your finances decreases, allowing you to focus more on strategic financial planning rather than just keeping up with the administrative tasks.

Additionally, fewer financial accounts can lead to enhanced security. Each account you hold potentially exposes you to risks like identity theft and other fraud. This is especially true if they are accounts you infrequently monitor. Reducing the number of accounts minimizes these vulnerabilities, making it easier to secure your financial information. Having fewer accounts also simplifies the process of updating security settings and passwords, which increases your security.

Having fewer accounts also means there's less to think about when it comes to automation. Below are the steps you should take to reduce the number of accounts:

1. Assess usage and necessity, using the account list you created:

    a. For each account, note how often you use it and for what purpose.

    b. Identify any accounts that are rarely used or have been replaced by more efficient solutions.

2. Evaluate fees and benefits:

    a. Check whether any accounts have fees that are not justified by the benefits they offer, such as high maintenance fees or low interest rates.

    b. Consider the rewards or benefits associated with each account and whether you're actually utilizing them.

3. Consolidate accounts:

    a. Where possible, consolidate activities into fewer accounts. For example, merge savings from multiple banks into one with the best interest rate or benefits.

    b. Transfer credit balances from seldom-used cards to one with better rewards or lower interest rates, if possible.

4. Close the accounts you no longer use:

   a. Contact the financial institutions to close accounts that are no longer needed. Make sure to get confirmation in writing for your records.

   b. Note: Be mindful of any potential impacts on your credit score when closing credit accounts. Closing accounts can temporarily have a negative impact on your credit score.

5. Update automatic payments and direct deposits:

   a. Redirect any automatic payments or direct deposits linked to the accounts you're closing to your remaining accounts.

   b. Update any saved payment methods on online shopping sites, utility providers, or service subscriptions.

After you go through this exercise, your finances will be much easier to track, saving you time. In addition, you'll have reduced your identity theft and fraud footprint as well as made automating your finances much easier.

Automate Paying Your Bills

Now that you've consolidated and reduced your financial accounts, it is time to work on saving time with financial automation. We'll start with paying your bills.

Life involves paying a seemingly ridiculous number of bills. You have a rent/mortgage, taxes, utilities, credit cards, medical bills, car notes, and so much more. To further complicate matters, some bills can be regularly paid online, some can be paid with credit cards, and some accept only checks or direct debit. Also, some bills come monthly, some come quarterly, and some are completely irregular. Not to mention, some bills are the same amount each time, while others can vary greatly.

If you have felt like paying your bills takes a lot of time, or you have a hard time keeping up with your bills, you're not alone. Keeping track of all your bills can be a daunting task. Many bills are missed and paid late because people forget to pay them, not because they don't have the money to pay them. Luckily, there is a better way, and that way is to automate paying your bills.

The variability of payment frequencies, amounts, and methods makes automating your bill payments somewhat complicated, but it is still doable. Here are the steps:

1. **List all regular bills:**
   - Start by creating a list of all your recurring bills, such as utilities, rent or mortgage, credit cards, loans, insurance premiums, and subscriptions. Note the amount (if fixed), the due date, and the payment method currently used. If you did the cash flow exercise mentioned earlier in this book, then you should have most of this information already.

2. **Choose a payment method:**
   - For each bill, decide whether you want to automate payments from your checking account, through a credit card, or from a digital wallet. Each has its advantages:
     - Checking account: Direct debits are reliable for fixed bills like mortgages, and usually there are no processing fees.

- Credit card: Good for earning rewards and improving credit scores, provided you pay off the balance each month.

- Bill pay services: Helps you avoid fees and allows you to automatically pay bills that don't have their own payment portals.

- Digital wallets (e.g., Apple Pay, Google Wallet, PayPal): Useful for smaller or variable payments.

3. **Set up payments directly with billers:**

   ○ Visit each biller's website and log in to your account.

   ○ Navigate to the payment section and select the option to set up automatic payments.

   ○ Enter your preferred payment method details and authorize the biller to charge this account on a recurring basis.

   ○ Verify each setup by confirming the start date and frequency of payments.

4. **Set up automated payments through your bank (bill pay services tied to your banking account for bills you can't pay directly with the provider):**

   ○ Log in to your online banking platform.

   ○ Look for a section titled "Bill Pay" or "Payments."

   ○ Add each biller's information, such as the company's name, your account number with them, and their payment address or electronic payment details.

- Set up recurring payments specifying the amount, the date to send the payment, and the frequency (e.g., monthly, quarterly).

5. **Monitor and adjust:**
    - After setting up automatic payments, monitor your bank statements and bill statements for the first few months to ensure everything is processing correctly.
    - Keep an eye on any emails or messages from your billers in case there are notifications about payment issues or changes in the amount due.

6. **Maintain sufficient funds:**
    - Ensure there are always enough funds in your chosen payment account to cover the automated bills. This might involve setting up balance alerts with your bank.
    - Consider setting up a linked savings account or overdraft protection as a backup.

7. **Review regularly:**
    - Schedule an annual or bi-annual review of your automated payments. Check for any changes in your billing amounts, any new bills, or services you may have canceled but are still paying for.

Automate Saving and Investing

The key to reaching your intermediate and long-term financial goals is through accumulating savings and consistently investing. You need savings to pay for unforeseen short-term expenses like car and house repairs. You also need savings to pay for short-term planned expenditures like vacations and new car purchases. If you don't have sufficient savings for these expenditures, you'll need to use debt financing (e.g., credit cards, home equity loans, borrowing from friends), or you may have to sell investments at a time when they are down. This makes the expenditures more costly.

Investments are important for your long-term goals. Investing in the stock market is a proven way of accumulating wealth over time. This wealth accumulation can help you buy a home, pay for your children's college, retire, and more.

Saving and investing work best if they are habits. Don't wait until you get around to it each month or quarter or year. You'll always find something else to spend the money on that you should have saved or invested, or you just won't get around to the task. That's human nature.

So get started with wealth-building by following these steps to automate your saving and investing:

- **Step 1: Set clear financial goals:** Do you want to save for a vacation, a down payment on a house, or retirement? For instance, if you're looking to save $30,000 for a home down payment in three years, you'll need to set monthly savings targets.

- **Step 2: Choose the right savings account:** Open a high-yield savings account or a dedicated savings account with a bank or credit union that offers good interest rates. A seemingly small difference in interest rates can make a substantial impact on your savings over time.

- **Step 3: Set up automatic transfers:** Most banks allow you to schedule automatic transfers from your checking account to your savings account. If your checking account where you deposit your paycheck is at the same firm as your savings account, you likely can set up automatic transfers. Many banks and other financial institutions also have a way of transferring money to other institutions, using routing and account numbers. Next, decide on a fixed amount to transfer each month, based on your goals. For instance, if you want to reach a $10,000 goal in three years, you could set up a monthly transfer of approximately $278 to your savings account.

- **Step 4: Automate retirement contributions:** If you have an employer-sponsored retirement plan, like a 401(k), enroll in it and set your contributions to be automatically deducted from your paycheck. Aim for at least the company match to maximize your savings. For example, if your employer offers a 50% match on contributions up to 6% of your salary, ensure you contribute at least that amount (6%).

- **Step 5: Open an investment account:** Consider opening an Individual Retirement Account (IRA) or a brokerage account for long-term investing. Research low-cost investment platforms like Vanguard or Fidelity, which offer automated investment options. Also, consider investing in a target-date fund that automatically reallocates your portfolio as you approach retirement or whatever goal(s) you have.

- **Step 6: Set up automatic investments:** Just like with your savings, set up automatic contributions to your investment account. For example, you could choose to invest $200 each month into a diversified index fund. Over time, this can lead to significant growth through the power of compounding interest. The main thing is to establish the habit and reduce the need to make an active decision each month to invest. Put it on autopilot.

- **Step 7: Monitor and adjust:** Even with your investing and saving on autopilot, you still need to monitor your progress and make adjustments as needed. Perhaps your goals have changed. Maybe you are making more money than you used to. Or, you could decide on a different investment strategy. These changes may spur you to adjust your plans.

Regularly review your progress toward your financial goals. Many investment platforms provide insights and tools to help you track your investments. For example, if you find that you're consistently exceeding your savings goal, consider increasing your planned monthly contributions or adjusting your investment strategy. Alternatively, you could decide to spend more money on other things you enjoy!

I recommend reviewing your savings and investment allocations and plans at least once every year, even if nothing spurs you to. Sometimes just the act of reviewing your finances will give you ideas, plus you always need to check that your plan is leading to your goals.

In the Real World: Automating Saving and Investing

Consider Sarah, who wants to save for a new car while also preparing for retirement. She sets a goal of saving $5,000 in two years for the car. She opens a high-yield savings account and sets up an automatic monthly transfer of $210. Simultaneously, she contributes 10% of her salary to her 401(k) and decides to automate an additional monthly investment of $150 into a diversified exchange-traded fund (ETF). By automating these processes, Sarah not only achieves her car savings goal but also builds a solid foundation for her retirement without needing to constantly manage her savings and investments.

Automate Accounting

You probably have multiple bank accounts, investment accounts, loan accounts, and other accounts with various institutions, making keeping track of your money more difficult.

You need to track your money across all your accounts so you know where your money is coming from and where it is going. Without accounting, you may not know that you are overspending on eating out, or that membership programs and services you joined years ago still charge you, even though you no longer use them. Also, without accounting, you may not know that a criminal is illegally siphoning money from your accounts or charging your credit cards.

In the past, most people relied on keeping an up-to-date checkbook to record this information. Manual accounting from a dozen or so accounts is likely to be time-consuming and error-prone. Most people don't even bother to track expenses in any type of detailed manner for this reason. As I wrote earlier, it's important to track your money for planning and security reasons. Having all these accounts also makes automating accounting even more important. Here are three ways to automate your accounting:

Hire an Accountant or Bookkeeper

When people talk about hiring an accountant, they mean a lot of different things. They may be referring to a Certified Public Accountant (CPA), a bookkeeper, or even a tax adviser. For the purposes of the accounting activities you should automate, I'm going to describe what accountants and bookkeepers do.

Accountants typically have specific training and a college degree in accounting and can handle bookkeeping chores. They tend to cost more than bookkeepers, especially if they are a Certified Public Accountant (CPA). Accountants can do the services bookkeepers can do, but they also usually provide consultation and analysis. For example, they may give you advice on how to reduce your expenses, and other financial strategies for maximizing your money. They may also give you tax advice. Some accountants can also do your taxes for you.

Bookkeepers provide concierge services for your personal finances. They can handle paying your bills, balancing your checkbooks, and reviewing your credit card statements. Bookkeepers often work without specialized training or certification. However, they can obtain certification from organizations like the American Institute of Professional Bookkeepers or the National Association of Certified Public Bookkeepers.

If you are looking to simply automate your financial accounting, a good bookkeeper is all you need. If you want financial advice and tax assistance, you should consider an accountant, but it will cost you more. Either way, you'll need to give the bookkeeper or accountant access to your financial accounts and receipts so that they can do their job properly. You also should be prepared to meet with them regularly to go over your finances.

Use Accounting Software

You can use accounting software if you prefer a more hands-on approach to automating your finances. There are many personal finance and accounting apps and services like Quicken, YNAB (You Need a Budget), and Rocket Money, which can streamline your accounting processes. Most of those tools do much more than basic accounting. Also, many banks, brokerage firms, and other financial institutions offer accounting and personal finance software.

Choosing the right tool for your accounting and financial tracking needs is important. You'll want to make sure the tool matches what you want to use it for. It's also important that the tool can sync with your primary financial accounts so you don't have to do a lot of manual entry. Below is a list of features that you may be interested in:

- **Budgeting tools:** detailed income and expense-tracking capabilities
- **Expense categorization:** categorization of transactions, preferably automatically
- **Bill pay features:** ability to send payments to companies and reminders of bill due dates
- **Goal-setting:** setting and tracking of financial goals
- **Investment and debt-tracking:** features that can track your investment (e.g., 401(k), brokerage, IRA) and debt (e.g., credit cards, mortgage, student loans) accounts

- **Real-time syncing**: automatically updating your accounting with transactions from your financial accounts in real time

- **Reports and analytics**: detailed reports that illustrate your financial status and provide insights into areas of improvement

- **Security**: good encryption capabilities and additional measures like two-factor authentication

- **Platform availability**: access to the software on your preferred platforms (e.g., iOS, Android, macOS, web browser)

When you are evaluating tools, be sure to choose one with the features you like. Automating your accounting with one of these tools can save a lot of time and give you valuable insight into how your money flows and how you can retain more of your earnings.

Combine Accounting Software with an Accountant or Bookkeeper

Another approach is to combine accounting software with an accountant or bookkeeper. You could choose an accounting professional who is knowledgeable in the software you want to use. They can help get you set up and started. Periodically (e.g., quarterly or yearly), they can look over your work and see whether there are any issues with how you are doing accounting and/or point out places where you can improve your finances.

Automate Giving and Sharing

Some people donate to various charities. Some people support their adult children financially. Also, some people provide financial assistance to their parents and friends. I'm here not to judge—just to show you a few ways to automate your giving.

Charities

Automating your charitable giving is an effective way to ensure that your contributions are consistent and aligned with your values, all while simplifying the process. Automatic donations simplify supporting causes you care about without having to remember to give each time. Fortunately, automating charitable donations is usually straightforward.

Begin by selecting the charities or causes you wish to support. These could include local nonprofits, international organizations, or specific initiatives that resonate with your values. Research these organizations to ensure they are reputable and align with your philanthropic goals.

Next, determine how much you can afford to donate. Establishing a budget helps you make informed decisions about how much to allocate to each charity. For example, if you decide to donate $1,200 a year, you could opt to give $100 each month to your selected charities.

There are a few ways to automate giving to charities. Most charitable organizations offer options for automated donations. You can set up recurring contributions directly through their websites. Look for a "Donate" button, which allows you to select a recurring option—monthly, quarterly, or annually.

Another way to automate donations is by using platforms that facilitate automated giving to multiple charities like JustGiving and GoFundMe. These platforms often provide features for setting up recurring donations to various causes in one place, making it convenient to manage your giving.

Make sure to track your charitable donations no matter which method you use. Many organizations send annual statements, but keeping your own records can help you stay organized and ensure you meet your giving goals. This practice is also beneficial for tax purposes, as charitable contributions are often tax-deductible.

Automating Financial Support for Friends and Family

Sometimes friends and family will ask for some money here and there. Those situations probably don't need automation, as giving cash, writing a check, using Venmo, etc., will probably work in a pinch. However, sometimes we have loved ones that we decide to support regularly, like giving an allowance to an adult child in college or helping a parent with medical costs. It can become tedious to manually mail a check or initiate a Venmo transfer every month, or even every couple of weeks. Fortunately, there are good ways to automate money transfers to people too.

Automatic Transfers

Most banks will allow you to set up automatic transfers between accounts and sometimes even with other financial institutions. In most cases, however, you will need to be a named owner of the account. This works well for children. As an example, I set up joint accounts at a local credit union with my children when they were young. I also have a checking account at this credit union. I can, in real time, put money in their accounts, and I can also set up scheduled transfers from my checking account to theirs.

Peer-to-Peer Apps

There are situations when you can't or don't want to be a co-owner of an account. In those cases, you can use a peer-to-peer payment app like Zelle. In fact, many banks have partnered with Zelle to offer this service. With Zelle, you need to know only the person's email address or mobile phone number, and you can send them money directly from your bank account. Zelle transfers are instant. Venmo is another peer-to-peer app, which can do automatic transfers.

Bill-Paying Services

A third method is to use a bill-paying service to send a check every month to an individual. Many banks and credit unions offer bill-paying services for free. You would need to know the individual's name and mailing address, and the recipient would likely have to deposit a check sent by the service. These days, you can deposit checks using a mobile app on your phone. It's made depositing checks more convenient than it used to be, but one of the aforementioned methods is preferable due to speed of transfer and simplicity.

Paying Bills Directly

Sometimes, financial support of loved ones involves directly paying rent, tuition, or other bills. In this case, you can set up automatic deductions or charges to your credit card directly with the services, just as you would if you set it up for yourself (as mentioned in the Automate Paying Your Bills section earlier in this book).

Systematize Analysis and Decision-Making

Automation greatly simplifies bill-paying and transferring money, but how can it help with more complex money management decisions? For example, creating, adjusting, and monitoring your budget can take a lot of thought. It can be hard to know which adjustments to your budget are most realistic and useful, where you might be consistently over or underspending, and how or if you should adjust your budget to account for changes.

Automating personal financial decisions and analysis may seem scary. However, you can automate at least parts of the process, making it easier to stay on track with your goals while minimizing the stress of everyday money management. Here are a few effective ways to systematize and automate your personal financial decision-making.

Utilizing AI for Financial Analysis

AI technology has advanced to the point where it can analyze your financial behavior and offer personalized recommendations. For instance, accounting apps like Simplifi and Rocket Money utilize machine-learning algorithms to categorize your transactions, identify spending trends, and suggest adjustments to your budget. These tools can alert you when you exceed your spending limits or when recurring expenses increase unexpectedly, empowering you to make proactive decisions about your finances. Additionally, AI can help simulate different financial scenarios, such as the impact of a significant purchase on your overall budget, allowing you to assess potential outcomes before making decisions.

Setting Purchase Thresholds for Couples

For couples, financial discussions can sometimes lead to tension, especially when it comes to spending decisions. One effective strategy to minimize friction is to establish predefined thresholds for purchases that can be made without needing to consult each other. For example, you might agree that purchases under a certain amount, say $100, can be made without prior discussion, while purchases above that amount require mutual agreement. By automating reminders or alerts when a budget category nears its threshold—perhaps through a shared budgeting app—couples can maintain transparency and communication around their finances without the burden of constant deliberation.

Establishing and Adhering to Budgets

Creating and sticking to a budget can be challenging, but automation can simplify this process. Start by using budgeting software that automatically tracks your income and expenses, categorizing them in real time. Then, you can set up a monthly budget that allocates specific amounts for categories like groceries, entertainment, and savings. AI-driven apps can also analyze your historical spending data to provide more accurate budget recommendations, adjusting automatically based on your lifestyle changes. To ensure adherence to your budget, enable notifications for when you approach your limits in specific categories. This proactive approach keeps you informed and accountable, making it easier to adjust your spending behavior in real time. Alternatively, you could use an accountant or bookkeeper to monitor your spending versus your budget.

Regular Financial Check-Ins

You can do a lot with financial automation, as we've seen, but it's still not clever enough to run your household accounts with no input. You'll still need to check in occasionally to keep an eye on your financial health. Most tools have features that can help you automate financial check-ins effectively without spending a lot of time.

Schedule automatic reports or summaries from your budgeting apps or accounting professional, which highlight your spending habits, savings progress, and overall financial situation. For example, a weekly or monthly digest can provide insights into how well you're adhering to your budget and whether you're on track to meet your savings goals. These reports can also highlight areas for improvement, such as overspending in certain categories, allowing you to adjust your financial strategy as needed.

What I Do

Personal finance is one of my passions. I spent part of my career as a personal financial advisor. So I have spent a good amount of my lifetime improving and automating my finances. Here are some of the things I've done:

Pared My Financial Accounts

Early in my adulthood, I amassed too many different bank accounts. I had a bank account that I used to keep money from my high school jobs, a bank account opened in college so I could access banking services close to me, and bank accounts opened as a result of working summer jobs in different places. I also amassed various credit cards over that time. Not to mention, when I got married, between the two of us we had duplicates of everything.

Additionally, as I changed jobs, I ended up with multiple workplace retirement accounts (e.g., 401(k) and 403(b) accounts). My wife and I had IRA and brokerage accounts at different institutions.

All these accounts made it hard to keep track of where my money was. It also meant monitoring multiple accounts for fraud, making multiple credit card payments every month, not to mention keeping track of changes to the accounts including finance fees, interest rates, investments, and credit limits. It simply wasn't worth having all these accounts.

Over time, we pared them all down to one personal bank account for my wife, one for me, and a joint bank account that bills are paid out of. We have one Visa, one Mastercard, and one American Express card, which we share. I've moved workplace retirement accounts to either my current workplace or a single financial institution that holds my other IRAs and brokerage account. This is still a lot, but it is much simpler and less time-intensive to track than having so many different accounts.

Automated Transfers to Savings, Investments, and Family Members

Consolidating these accounts has made it much easier to put our investing and savings on autopilot. Aside from the paycheck deductions that go into our retirement accounts, we also have automated monthly transfers into our savings accounts and directly into investments like our children's 529 college savings accounts. Additionally, we have automated transfers from our bank accounts to family members, such as allowances that we give to our children.

Tracked Spending with an App

I use an app called Monarch Money to automate my budget-reporting and track my spending. I've connected Monarch Money to all of my financial accounts. I've set it up to automatically categorize my expenditures so I can manage our expenditures to our budget. Its reporting features help me get a good idea of where my money is going and allows me to adjust my spending when desired. This is so much easier than keeping track of receipts and manually tracking my expenditures while using a spreadsheet.

Scheduled Regular Reviews

I have a recurring task of reviewing my overall financial situation once each year in January. At this time, I review how my finances have changed over the year and what my progress is toward my financial goals, along with any changes I need to make. I compile all this information into a set of charts and graphs to make them easy to digest for conversations with my wife. It is nice to see how far we've come. Not so nice to see how far we have to go toward our goals, but it is better to know and plan!

This may seem like overkill, but I also give financial presentations to my children about their college funds each year. And, I teach them the value of saving by giving them a 50% match on what they have in their savings account at the end of the year.

Plans for the Future, Using My Own Expertise, Financial Planner, and AI

Because I have a professional and academic background in financial planning, I do most of it by myself. That said, as I near retirement, I plan on working with a retirement-planning specialist to optimize and double-check my plans and progress. I also have frequent "conversations" with AI chatbots like Gemini and ChatGPT about how to best invest, how much money to save for retirement, and how to calculate different scenarios.

It's perhaps not surprising that computers excel (pun intended) at accounting. After all, the first spreadsheet was invented back in 1979. In the next chapter, we'll explore a more personal arena—your health and fitness.

Key Takeaways

- **Simplifying accounts**: Consolidate financial accounts to reduce complexity and make it easier to manage and automate your finances.

- **Cash flow review**: Regularly review your income and expenses to identify areas for automation, such as recurring bills and subscriptions.

- **Automating bill payments**: Set up automatic payments for regular bills to reduce time spent managing finances and avoid late fees.

- **Saving and investing**: Set up automatic transfers to savings and investment accounts, ensuring that you're consistently working toward your financial goals without having to think about it.

CHAPTER 5

AUTOMATING YOUR HEALTH AND FITNESS

Discipline is remembering what you want most.
Not what you want now.
—Billy Blanks

Health and fitness are the foundation of your life's happiness. It's hard to enjoy many aspects of life without your health. Good health and fitness also allow you to enjoy a longer life, so it's important to invest time and effort into your own physical well-being. And always remember that good health and fitness are not gifts that someone can give you; you have to make the effort.

Think of your desired health and fitness as a destination, and getting there a journey. Like on any long-term journey, there will be obstacles and setbacks, but you must persevere. With health and fitness, the journey is lifelong. Even if you reach the level of health and fitness you want, you still have work to do to keep yourself there.

How can you ensure that you stay on this path your whole life? The answer is by automating as much as possible, removing barriers to your success to improve your results. Exercising regularly, eating well, and all the other disciplines that come with trying to be healthy take a lot of work. So it's important to use automation tools that help you focus on the work and the results, not the minutiae.

Set Your Health and Fitness Goals

Setting health and fitness goals is a crucial first step in creating a sustainable wellness journey. Goals provide direction, motivation, and a clear benchmark for measuring progress. When defining your health and fitness goals, it's essential to make them SMART—Specific, Measurable, Achievable, Relevant, and Time-bound. For example, instead of aiming to "get fit," a SMART goal might be to "run a 5K in under 30 minutes within three months." This approach not only clarifies what you want to achieve but also establishes a timeline and metrics for success.

When it comes to health and fitness, there are many types of goals you might consider setting. Here's a list to get you started:

- **Weight loss:** Aim to lose a specific number of pounds or reach a target body fat percentage. Tracking this helps you stay motivated and adjust your eating and exercise habits accordingly.

- **Strength training:** Set goals for lifting specific weights or completing a certain number of repetitions in key exercises (like squats, bench presses, or deadlifts). Tracking your progress helps you see gains and keep pushing yourself.

- **Cardiovascular fitness:** Aim to improve your endurance by running a certain distance or completing a specific time for a cardio workout. Monitoring your heart rate during workouts can help gauge your improvement and intensity.

- **Flexibility:** Work toward achieving specific yoga poses or increasing your range of motion in certain stretches. Improving your flexibility can enhance overall fitness and reduce injury risks.

- **Nutrition:** Set goals for daily caloric intake or specific macronutrient ratios (carbs, proteins, fats). Tracking your meals can help you stay aligned with your dietary goals and make healthier choices.

- **Mental health:** Incorporate goals related to mindfulness or stress reduction, such as meditating for a certain number of minutes each day. Tracking your mood or stress levels can enhance your overall well-being.

Once you have your goals in place, tracking your progress toward them in an automated way can make the process not only easier but also more enjoyable! Fitness apps like MyFitnessPal or Lose It! are fantastic for logging workouts and meals effortlessly. These apps often feature built-in reminders, so you can set daily alerts to log your food intake or workout sessions—making it feel like less of a chore and more of a routine.

Wearable technology, like smartwatches or fitness trackers, can be a game changer. These fabulous devices provide real-time feedback on your activity levels, heart rate, and even your sleep patterns. For example, if your goal is to hit 10,000 steps a day, your fitness tracker will not only count your steps but will send you gentle nudges if you're falling behind, keeping you accountable without the added stress.

Additionally, automating your meal planning and grocery shopping can complement your fitness goals beautifully. Use meal prep apps that create grocery lists based on the recipes you select for the week. Some services even offer delivery, saving you time and ensuring you have all the healthy ingredients on hand to support your goals. We'll discuss automating meal planning in more detail later in this book.

What I Do

Health and fitness have been an important part of my adult life. I've run marathons, participated in sports leagues, and am generally an active person. However, my journey has not been linear. I have struggled with fitness and weight, especially as life has sometimes thrown curveballs (e.g., career changes, the birth and raising of children, death of loved ones, injuries).

The set and automated tracking of my goals has served me well throughout the twists and turns of life. Here are two examples of those twists and turns and the way technology and automation helped me reach my goals:

Graduate School at Night

In my mid-twenties, I went back to graduate school. However, I wasn't in a financial situation where I wanted to quit my job and go back to school. So instead, I found a program that had night classes. For three years, I worked during the day and took classes at night and on the weekends. I didn't have a lot of free time, and for the first time in my life I got out of good fitness and became overweight.

When I finished grad school, I decided it was time to focus on getting into good shape and set the following goals:

- Work out five times a week for a year
- Reduce my body fat percentage by at least 10% in a year
- Run a 5K in under 25 minutes

I planned and documented my workouts in a spreadsheet (this was before smartphones and apps!). I also took weekly body measurements to calculate my progress in reducing body fat and logged those in a spreadsheet. These days, you can buy a smart scale that will calculate and track your body fat percentage for you.

Now, I had never been a runner before, so the last goal was definitely aggressive. To do that, I planned my running workouts long in advance and got a sports watch that would track my estimated distance. I could then transfer that data to a program that tracked all my runs. It was pretty advanced at the time. All I had to do was sync the watch with the program on my computer.

Tracking these goals played a large part in my success in achieving them. Seeing continual progress each week was encouraging. I ended up getting in great shape and surpassed my running goal by far. Upon discovering I loved running, I went on to run much faster than a 25-minute 5K and even ended up running marathons!

Tearing My Achilles

A few years back, not realizing that in my 40s I could never be Superman, I tore my Achilles. I was in pretty good shape at the time. I'd gone for a seven-mile jog in the morning and then went to my 40-and-up basketball league to play at night. I was having a great game, feeling good and then... snap! I tore my Achilles. Then, partly due to inactivity and partly due to depression caused by knowing I had a long road to recovery ahead, I gained weight and lost fitness.

But as I worked through rehab, I was determined to regain my fitness. I set a goal of not only being able to run again but to run a half-marathon in a little over a year. I set running goals and another goal to lose the 20 pounds I had gained.

Automation helped with tracking both goals. I put together a running plan and tracked all my runs using my Polar smartwatch. This watch kept track of speed, distance, and heart rate as I ran, and automatically uploaded the results so I didn't have to track my progress manually. I also purchased a smart scale, which automatically logged and charted my weight during each weighing session. (As I alluded to earlier, the scale also measures and tracks body fat percentage.)

Getting my runs to be faster, longer, and more efficient over time, as tracked by Polar, along with watching the line of my weight slope downward was really motivating. I didn't have to remember to write anything down. I just had to put in the work and observe the results. It may have been the slowest half-marathon I ever ran, but I still completed it the year after I had torn my Achilles.

Set a Workout Schedule

Some people love working out; others absolutely abhor it. I'm somewhere in the middle. I love the fitness progression, which working out gives you. I like being able to eventually run farther, run faster, lift more, etc. But I don't love every single workout.

You'll benefit from setting a workout schedule no matter where you fall in the spectrum of workout love. Scheduling your workouts in advance allows you to put your workouts in "set it and forget it" mode. No more going to the gym and wondering what workout you should do. If you schedule them, you'll always know in advance what workout you will be doing, and you'll be able to mentally prepare for it.

Scheduling your workouts not only saves you time and mental energy from figuring out what you're going to do during each workout; it also makes your workouts more effective. I know from experience and research that it's important that your individual workouts fit together in a program.

Let's take running, for example. Hard running days—i.e., days where you are running for long periods or doing draining interval runs—should be followed by recovery runs where you are running easier. If you stack your hard runs over and over, you will burn out, get injured, or both. Scheduling your workouts allows you to avoid this scenario.

This same principle applies to more than running. When it comes to weightlifting, you probably don't want to work the same muscles hard two days in a row. Or, you may not want to work out your legs the day before you plan to do a long bike ride. Your workouts will benefit you more if they are scheduled and coordinated.

What's a good workout schedule? It's one that progresses you toward your health and fitness goals. It's one that clearly tells you what exercise you're going to do and when you are going to do it. It's one that assures that the workouts complement each other. You can write down your workout on a paper calendar, in a Google calendar, or in some type of workout app.

I like to put my workouts in an online calendar. I tend to plan 8–12 weeks of workouts in advance. Sometimes I make adjustments to the schedule, but I always have a schedule. I do most of my workouts first thing in the morning. Thus, I don't have to think about it during early morning grogginess.

Get a Fitness Tracker

Setting goals and creating a workout schedule are important for your fitness journey, but they are not sufficient. Of course, you have to do the workouts. It's also important to track the way your workouts go and your progress toward your short-, medium-, and long-term goals. You may think you are working out hard, but perhaps you aren't making enough progress. It could be that you aren't working out hard enough or that the rest of your time spent each day is too sedentary. Or maybe you are constantly getting injured and feeling tired because you are working out too hard!

Fitness trackers are an essential tool for automating your fitness journey. These devices not only track your workouts but also monitor your overall activity, providing real-time insights into your progress toward your health and fitness goals. A fitness tracker can help you stay on track and optimize your routine toward endurance, weight loss, building strength, and other fitness goals.

When you use a fitness tracker during workouts, you can track a range of metrics such as heart rate, calories burned, distance, steps, and even specific workout stats like pace and duration. This data gives you a clear picture of your exercise intensity and performance. For example, if you're training for a 5K, a fitness tracker can help you monitor your running pace over time. Analyzing this data allows you to see improvements and identify areas where you may need to push harder or adjust your technique. If your goal is to run faster, tracking your pace and comparing it over several weeks will help you understand whether your training is effective enough or you need to increase the intensity of your sessions.

The fact that fitness trackers monitor your general activity in addition to your workouts is important because general activity plays a significant role in your overall health and fitness in addition to individual workouts. For example, walking or taking the stairs may seem like small actions, but they add up over the course of a day, contributing to better cardiovascular health, improved metabolism, and overall energy levels. Fitness trackers can remind you to stay active even when you're not doing a formal workout, ensuring that you're getting enough activity throughout the day.

You can fine-tune your workouts and make adjustments to achieve your goals more efficiently by analyzing the data from your fitness tracker. If your tracker shows that you're consistently meeting your daily step goal but falling short on your calorie burn, it may suggest that you need to ramp up the intensity of your workouts. Similarly, if you notice a plateau in your running pace, you can adjust your training program by incorporating interval training or increasing your distance incrementally. Some fitness trackers also help you avoid overtraining or injury by tracking your recovery and providing data on sleep quality, heart rate variability (HRV), and rest periods. This makes it easier to listen to your body and adjust your workout schedule to ensure you're progressing without burning out.

Incorporating the data from your fitness tracker into your fitness plan doesn't just make your workouts more effective—it also helps you stay motivated. Seeing the numbers on your tracker improve week after week can keep you engaged in the process, making it clear that your efforts are paying off. Plus, with many trackers syncing to apps that provide detailed reports and visual progress charts, you can easily track your achievements and adjust your goals as needed.

As I wrote earlier, I currently use a Polar multisport watch as my fitness tracker. However, there are many other trackers on the market. Garmin and Google/Fitbit are other major players in this area. And if you want a smartwatch, the Apple Watch, Pixel Watch, and Samsung Galaxy watches all have excellent fitness tracking capabilities. The key is to define the features you're looking for in a fitness tracker, including how it looks, how long the battery lasts, and how much it costs. There are many to choose from, and the right one for you is out there!

In the Real World: Automating Health and Fitness

James is a busy professional who struggles to stay on top of his health goals. He invests in a Fitbit Charge 6 to track his steps, heart rate, and sleep quality. The Fitbit syncs with the Fitbit app, which not only logs his data but also offers personalized workout and wellness recommendations. He sets daily reminders to move and tracks his progress on his phone. The phone app adjusts his exercise routine based on his current fitness level, ensuring that he's staying active but not overexerting himself. By automating his health-tracking with Fitbit, James stays motivated and maintains a healthier lifestyle without having to think much about it.

Schedule Doctor Visits in Advance

One of the easiest ways to stay on top of your health is by ensuring you never miss a doctor's appointment, especially those critical wellness checks that help detect potential issues before they become serious. Automating your doctor visits, especially routine checkups, can save you time, reduce stress, and ensure you're consistently monitoring your health.

The first step in automating your doctor visits is to schedule your annual wellness checkups well in advance. Many healthcare providers allow you to book appointments a year ahead, which means you can plan your appointments for the following year. Lock in your next visit before you leave the office so it's already on your calendar and won't be forgotten or neglected. This is especially useful for annual physicals, eye exams, dental cleanings, and other routine checkups that are easy to overlook.

Once your appointments are scheduled, most healthcare systems and doctors' offices offer digital reminders via email, text, or app notifications. Many patient portals let you schedule, reschedule, or even cancel appointments directly from your phone or computer, making the process seamless and easy to manage. You can set up automated reminders that alert you several weeks in advance, giving you plenty of time to adjust your schedule if necessary. This way, there's no scrambling to find an open appointment slot or forgetting about important wellness checks that could prevent serious health issues down the line. These services allow you to review test results and follow-up instructions, making it easier to stay on top of your health without extra effort.

Beyond scheduling, you can further automate your doctor visits by proactively managing prescriptions and follow-up care. Many health providers allow automatic refills for medications, so you don't have to worry about running out or remembering to call for prescriptions. In many cases, you can even have your filled prescriptions mailed to you. You can also set reminders for any follow-up appointments, vaccinations, or screenings that may be needed after your wellness check. These small automations help you maintain a consistent healthcare routine.

Put Self-Care on Autopilot

Self-care is essential for maintaining mental, physical, and emotional well-being. However, with busy schedules, it's easy to let it slip through the cracks. Automating certain aspects of your self-care routine can ensure that taking time for yourself becomes a regular part of your life. Automation can make scheduling regular pampering sessions and carving out moments for mindfulness both accessible and consistent, and with much less effort.

Regular Pampering: Manicures, Pedicures, and Massages

Taking care of your body isn't just about exercise and eating well—it's also about feeling good in your own skin. Automating appointments for manicures, pedicures, and massages is an easy way to incorporate regular pampering into your routine. Many spas offer subscription packages or memberships that allow you to schedule recurring visits automatically, ensuring that you get your self-care sessions on the calendar before life gets too busy. For example, you might set up a monthly manicure and pedicure appointment or book a bi-weekly massage to help ease muscle tension and relieve stress. Automation ensures that these moments of relaxation are not only prioritized but also built into your routine, saving you from having to find time for them later, which most of us won't do!

Meditation and Mindfulness Practices

One of the best ways to automate mindfulness is through apps and daily reminders. There are plenty of meditation apps, like Calm, Headspace, or Insight Timer, that can guide you through short sessions to clear your mind and reduce stress. Setting a daily reminder to meditate—whether it's before bed or first thing in the morning—takes the decision-making out of it. These apps even offer timed sessions that help you build a habit and track your progress over time.

Nature Walks and Alone Time

Connecting with nature and spending time alone are vital for recharging your emotional battery. Walking through the park, hiking in the woods or mountains, or simply sitting outside and breathing in fresh air are great ways in which nature provides a restorative space to clear your mind. Automating this aspect of your self-care could be as simple as scheduling a daily walk into your calendar or setting an automated reminder to step outside for a few minutes each day. Many people also find it helpful to plan a weekly "alone time" ritual—perhaps an hour of quiet reading, journaling, or a hobby you love. You're more likely to follow through with these activities if you schedule and set reminders for this personal time.

Other Popular Self-Care Options

Self-care goes beyond pampering and mindfulness. There are plenty of other activities that can nurture your well-being. For example, you might want to automate regular appointments with a therapist or coach, whether for mental health support, life coaching, or simply to check in on your emotional health.

Other self-care habits might include setting aside time to read or enjoy a hobby like knitting or painting, which can be easily overlooked in busy schedules. Setting reminders or even automating deliveries for books, crafting supplies, or art materials can help keep your hobbies alive, ensuring you prioritize these moments of joy and relaxation.

Even while you indulge in well-deserved R&R, there are countless small tasks around the home that you do without even thinking. Automating these tasks can free up valuable time—something we'll explore in the next chapter.

Key Takeaways

- **Health is foundational**: Good health and fitness are the basis for enjoying life fully, so prioritize creating consistent routines.

- **Automated fitness routines**: Use technology to track workouts, monitor progress, and set reminders, making it easier to stay consistent with your health goals.

- **Health data integration**: Connect fitness apps, wearables, and health-monitoring devices to automatically collect and analyze your data for better insights into your fitness progress.

- **Simplify health management:** Automate scheduling of medical appointments and health-related tasks to ensure your wellness needs are always met.

CHAPTER 6

AUTOMATING YOUR HOME ENVIRONMENT

A smart home is a home that works for you,
not the other way around.
—*Marja Koopmans*

You waste a lot of time without even knowing it. Turning lights on and off, adjusting your air conditioning, and setting the alarm every time you leave the house or go to bed. And that's before you get into larger chores like cleaning the floor. Each individual action is small, but they add up to hundreds of hours a year in lost time. And yet today, technology has given us a way of getting that time back.

Thanks to smart home technology, lighting adapts to your daily routine. Air conditioning adjusts the temperature by itself. And appliances like robot vacuum cleaners get on with chores by themselves.

A smart home is a residence equipped with technology that allows for the automation and remote control of various household systems and devices. Smart homes can be condos, houses, duplexes, townhomes, or apartments. They can be owned or leased. While they have different restrictions and capabilities, any type of home can be made to be smart. All that matters is that the devices are connected to the internet and can be managed on your smartphone and other devices.

A smart speaker gives you voice control over devices such as smart lighting and smart switches. Other examples of smart home devices are smart door locks, smart thermostats, and smart appliances, which can work on schedules, be controlled remotely, and be triggered by other things that happen. For example, a smart thermostat might cool a hot home when it detects that someone is inside. Smart lights may come on when it gets dark outside or when someone enters a room.

Home automation is very closely related to smart homes. It refers to the use of technology to control and manage household systems and devices remotely or automatically. With the help of smart home devices, you can automate everyday tasks like adjusting your thermostat, controlling lights, managing security systems, or even running appliances like coffee makers and robotic vacuums. This technology connects devices through a central platform, smartphone and tablet apps, and/or voice assistants like Amazon Alexa, Google Assistant, or Apple Siri. The beauty of home automation lies in its ability to make your living space more efficient, comfortable, and secure—all while saving time and energy.

One of the biggest advantages of home automation is the convenience it provides. The rest of this chapter is dedicated to ways to use smart homes and home automation to optimize your time and energy usage, and save money.

Automating Your HVAC System

HVAC, which stands for heating, ventilation, and air conditioning, has become one of the easier systems in your home to automate. Your home's HVAC system plays a critical role in maintaining comfort, but managing it manually can be time-consuming and inefficient. Smart thermostats and automated climate control devices are game-changers in this area.

The heart of automating your HVAC system is a smart thermostat. Devices like the Nest Thermostat, ecobee, or Honeywell Lyric can be controlled remotely through your smartphone, allowing you to adjust the temperature of your home from anywhere. These thermostats learn your preferences and adapt to your schedule over time. For instance, they can automatically lower the temperature when you leave for work, or raise it just before you return home.

These types of devices save you time and money. They eliminate your need to manually adjust the temperature settings each day. You no longer have to worry about whether you left the heat on when no one was home, or whether the settings are too hot or too cold. The thermostat does all the thinking for you.

Many smart thermostats integrate with voice assistants like Alexa or Google Assistant, so you can adjust your home's temperature with a simple voice command, saving even more time. Additionally, some systems are designed to track weather forecasts and make automatic adjustments based on outside conditions, further optimizing your indoor comfort without any effort on your part.

Another time-saving benefit is that many smart HVAC systems can alert you when it's time to perform regular maintenance, such as changing the air filter or scheduling an annual tuneup. Instead of waiting for a costly repair or experiencing uncomfortable breakdowns, you can get proactive reminders from your system to keep everything running smoothly. This not only saves time and hassle but can also extend the lifespan of your HVAC system and reduce long-term repair costs.

The energy-saving benefits of automating your HVAC system are significant. Smart thermostats are designed to reduce energy consumption by optimizing heating and cooling cycles. Instead of constantly running the heat or air conditioning when no one is home, the system can automatically adjust when you're away or asleep. This not only keeps your home comfortable when you need it but also prevents energy waste. Over time, these smart systems can help reduce your energy bills by up to 15%, according to some estimates, because they ensure that your system is working only when it's necessary. Also, many smart thermostats provide detailed energy reports, so you can see where you might be using energy inefficiently and make adjustments to further save money.

What I Do

If you hadn't guessed already, I'm really into smart homes. My book *The Smart Home Manual* is dedicated to helping people plan and build the smart home of their dreams. With that in mind, I'll share a few of the ways I've used smart devices and automation to help my HVAC systems run without intervention and save money.

Smart Thermostats and Combining My HVAC Systems

I use ecobee thermostats to control my primary heating HVAC system. It efficiently controls my heat and reminds me when I need to change air filters. However, I have two HVAC systems: (1) a hydronic heating system and (2) a heat pump-based system. They are unable to integrate with each other directly, but they both integrate with my central smart home controller, an application called Home Assistant.

Each system possesses unique characteristics. The hydronic system boasts exceptional energy efficiency, but it takes a considerable amount of time to heat a room from a cool temperature to a desired warm setting. Unfortunately, it lacks air conditioning capabilities. On the other hand, the heat pump system provides rapid heating but consumes more electricity than the hydronic system. Additionally, it offers air conditioning functionality.

To ensure efficient HVAC operations, I've implemented several automations that prevent simultaneous heating and cooling. I've also created automations that utilize the heat pump to quickly raise the temperature if, for example, the house has cooled overnight, while relying on the hydronic system to maintain the desired temperature. Thanks to my Home Assistant automations, these two HVAC systems seamlessly work together as a unified unit.

Additionally, I've taken the extra step of making my fireplace smart, allowing it to integrate with the overall HVAC setup. This intelligent combination ensures that all heating sources, including the fireplace, contribute to the optimal comfort and energy efficiency of my home.

Lastly, I also have a vacation mode for my home. When my home is put in vacation mode, the HVAC will turn on only if the house gets too hot or too cold. There's no need to keep heating and cooling to a comfortable level when no one is home.

Air Quality Monitoring

Since we primarily use hydronic heating around our house, we don't get much airflow without opening the windows. Unfortunately, we have a couple of people in my family who often get allergies. So, I decided to use air quality sensors and an air purifier to make sure that we know when we need to shut our windows. The air purifier automatically kicks off if the air inside our home becomes unhealthy.

This doesn't save us money, but it does make our life better without our having to do much of anything except for keeping the windows closed when we are alerted.

Bathroom Fans

When we first moved into this home, our kids would leave the bathroom fans on all the time. Although the fans are energy efficient, this is still a waste, and someone eventually has to turn them off. I installed timers in all but one of the bathrooms around the house. Now, you select how long you want the fan to be on, and then it automatically shuts off.

For the last bathroom, the bathroom my kids primarily use, I had a different bathroom fan problem to solve. My children often leave the fan on forever; however, they don't leave the light on when they leave the bathroom. So I have an automation that turns the fan off 20 minutes after the bathroom light turns off. It turns off the fan, no matter what, after an hour.

The only thing my kids forget to do more than turn the fan off is turn the fan on, which is important when they are taking their way-too-long showers. The bathroom fan always turns on if the lights have been on for five minutes.

I could also put a humidity sensor in the bathroom and have that be the trigger for the fan's turning on and off. That might be a future project.

Automating Your Home Lighting

Automating your home's lighting is a simple yet powerful way to improve your daily routine, save energy, and reduce costs. With a smart lighting system, you can control the lights in your home remotely, set them to turn on or off automatically, and even adjust their brightness or color based on your preferences—all without lifting a finger. This automation enhances convenience and also ensures your home is more energy-efficient, which can lead to significant savings over time.

One of the primary benefits of automating your home lighting is the sheer convenience it offers. With smart bulbs and lighting systems like Philips Hue, LIFX, or Nanoleaf, you can schedule your lights to turn on at specific times, such as in the morning when you wake up or as you're about to arrive home in the evening. Imagine coming home to a well-lit house without the need to fumble for switches or walk around the house turning lights on. You can even set up motion sensors to automatically turn on lights when someone enters a room and turn them off when the room is empty, saving you the trouble of constantly checking whether the lights are still on. Additionally, you can control your lights remotely through smartphone apps or voice commands via smart assistants.

Smart lighting also saves money. Traditional bulbs waste energy by staying on when not needed or by operating at full brightness all the time. By automating your lighting, you can ensure lights are on only when necessary. For example, if you forget to turn off a light when leaving the house, your smart lighting system can turn it off for you remotely, preventing wasted energy. Similarly, you can set up lighting schedules to dim lights or switch them off during certain hours (like at night when you're asleep).

Smart bulbs are usually energy-efficient LEDs, which use significantly less power than incandescent bulbs. By switching to smart LEDs, you can further reduce your electricity bill, as these bulbs consume a fraction of the energy of regular incandescent bulbs and last much longer—up to 25,000 hours or more.

Automated lighting can also improve the security of your home. With smart lights, you can set schedules or use geofencing to turn lights on and off when you're not home, giving the impression that someone is there. This simple feature can help deter potential burglars and provide peace of mind when you're away. Additionally, many smart lighting systems allow you to control your lights from anywhere in the world. When combined with surveillance cameras, you can check in on your home at any time and ensure everything looks secure.

Decorative lighting can add fun to your home while still saving you time. You can set up Christmas light automation that coordinates with the lights on your Christmas tree. You can even install year-round LED lights inside and outside your home that can be used for more than just Christmas. You could coordinate the colors with your favorite sports teams, special occasions, and other holidays.

What I Do

I've automated my home lighting in a lot of ways. First, all the lights in my house are either controlled by smart light switches and motion sensors or programmed on timers. For instance, the lights on the stairs to my basement turn on with motion and stay on for one minute past the moment when motion is no longer detected. My outdoor lights, however, are on timers; they turn on at sunset and turn off at sunrise for security purposes.

Most of my other lights are either smart bulbs or those controlled by a smart light switch. This allows me to do the following:

- Turn on and off lights, based on motion detection
- Turn off all the lights in the house when the alarm is armed away
  - Note: "Armed away" means that the alarm is armed, with the expectation that no one is in the house. This means that the alarm goes off if motion is detected. This is in contrast to when the alarm is armed "stay," which assumes people are in the home and only door and window sensors can set off the alarm.
- Turn on and off lights with voice commands
- Turn on lights when the alarm is set off to scare away intruders

I also have smart LED lights inside and outside my house. They are controlled by my home automation system. They change to different themes throughout the year, based on holidays (e.g., Independence Day, Memorial Day, and Easter) and seasons (e.g., spring, fall, and holiday/Christmas season). They also turn into an alert pattern when my alarm is set off.

Lastly, my house changes how my lights work when it goes into vacation mode. Indoor lights come on at strategic times of the day to simulate what would happen if we were at home. For instance, bedroom lights come on in the morning and at night.

Automating lighting has improved our home's security and aesthetics whilst boosting our mood and saving us time and money on decorations.

Alarm and Door Lock Automations

Investing in smart deadbolt locks allows you to better secure your home and ditch your keys. Smart deadbolts can be opened with a phone, a fingerprint, a code, or all three, and you can still get ones that work with a key if you'd like. What makes smart door locks good for security and saving time is that they can automatically lock.

You can set your smart locks to automatically lock at a certain time of night, so that even if you forget to lock your doors, they will eventually lock. You can also set them to automatically lock when you arm the alarm. Also, you can lock smart deadbolts from the outside without having a key, which makes things a little quicker.

Additionally, the smart door locks allow you to do the following:

- Remotely lock and unlock the doors
- Give someone a temporary guest code to enter your home
- Know who opened the door and when (for knowing when your kids get home safely)

Locking your smart deadbolts when the alarm is armed isn't the only alarm automation that can save you time and make your home more secure. You can tie your alarm into your smart home system so that it can do the following:

- Turn on lights in rooms where it has detected intruders

- Turn off all the lights in your home when the alarm is armed away

- Turn on a strobe light to disorient intruders

- Play an external siren if your alarm system's siren is weak or can be heard only inside or in part of the home

- Have your smart speakers tell intruders to leave and inform them that the authorities have been called

- Play loud and annoying noises from smart speakers

Additionally, you can have your alarm unlock your smart deadbolt when it is disarmed. I wouldn't do this with your front door. I have a smart deadbolt inside my garage, right next to the panel used to disarm the alarm. When I disarm the alarm, the garage door deadbolt unlocks so I can enter without having to unlock the door after disarming the alarm.

Automate Cleaning Your Floor with a Robot Vacuum

Cleaning your floors can be one of the most time-consuming chores, but with a robot vacuum, you can automate the process and reclaim your time for more important things. These clever little machines can navigate around your home, vacuuming up dust, debris, and even pet hair—all while you focus on other activities. Some advanced robot vacuums even include mopping functions, providing an all-in-one solution for floor cleaning.

The best way to take full advantage of a robot vacuum is by setting it on a regular cleaning schedule. Many robot vacuums allow you to program cleaning times, so your floors are automatically cleaned while you're at work, sleeping, or running errands. Simply set it and forget it—there's no need to worry about remembering to vacuum anymore. If your robot vacuum is equipped with advanced features like room-specific cleaning, it can be programmed to clean specific rooms at certain times. This means your kitchen and living room can get a thorough cleaning daily, while less trafficked areas are cleaned less frequently.

For homes with mixed floor types (like hardwood, carpet, or tile), some robot vacuums automatically adjust their suction power or cleaning method based on the surface. This leads to more effective cleaning across different floor types. Additionally, many models come with a mopping function, which is ideal for cleaning hard floors and removing stains. You can also buy a robot mop that works independently of your robot vacuum.

While robot vacuums are incredibly efficient, they aren't entirely foolproof. One challenge is the need to clear the floor of obstacles before cleaning. While most robot vacuums are designed to navigate around furniture and avoid large obstacles, they may struggle with small objects like cords, socks, or toys left on the floor. Some models have advanced sensors that help them detect and avoid these objects, but it's still a good idea to tidy up the floor before starting the robot vacuum, especially in rooms with lots of clutter.

Key Features to Enhance Robot Vacuum Use

To truly benefit from a robot vacuum, look for models with features that enhance convenience and efficiency:

- **Mapping and floor plan memory**: Many robot vacuums can remember your floor plan. They use sensors or cameras to map your home, allowing them to navigate more efficiently and clean systematically. This memory function lets them clean more intelligently, avoiding areas they've already cleaned or identifying spots that need extra attention.

- **Room-specific cleaning**: Some advanced robot vacuums can clean one room at a time or target specific zones in your home. This means you can schedule it to clean just the kitchen or focus on high-traffic areas like the living room, ensuring you get the best cleaning results where it matters most.

- **Cleaning on a schedule**: Automating cleaning schedules is one of the most convenient features. Set your robot vacuum to clean daily, weekly, or even multiple times per day, depending on your needs.

- **Self-emptying**: One of the most useful innovations in robot vacuums is the self-emptying feature. Some models come with a base station that automatically empties the vacuum's dustbin when it's full. This feature can save you time and effort, allowing the vacuum to run for longer periods without needing to be emptied manually, making it ideal for larger homes or those with pets.

- **Smart home integration**: Many robot vacuums are compatible with smart home systems like Amazon Alexa or Google Assistant. You can start, stop, or schedule cleaning sessions with simple voice commands, or integrate them with other smart home routines, like cleaning while you're out for the day.

I absolutely love my robot vacuum. It not only makes cleaning our floor simple but also does a better job than I do. I still occasionally mop our hard floors. However, whenever I sweep the floor, it never looks as good as when the robot vacuum cleans it. It gets all the crumbs and lint on the hardwood floors, almost making them look like they were mopped.

Using Smart Appliances to Automate Tasks and Save Time

Smart appliances are designed to make your life easier, save time, and boost efficiency. Let's talk about a few different types of smart appliances and the ways they can make your life easier.

Smart Fridges: Organizing and Streamlining Your Kitchen

A smart fridge is more than just a place to store food; it can help automate your kitchen. Many smart fridges come equipped with features like internal cameras that allow you to check the contents of your fridge remotely via an app. You no longer need to make a grocery list based on memory—simply open your app and see exactly what you have in stock, whether you're at the store or at work. Some smart fridges keep track of expiration dates, using a combination of AI and cameras to recognize the food type, manually entered expiration dates, and RFID tags. These fridges can send you reminders when items are about to go bad.

Additionally, some models allow you to add items to a shopping list by interacting with a screen built into the fridge or connect to online grocery services so that you can order items with a few taps. If you love cooking, certain smart fridges offer recipe suggestions based on the ingredients you have, helping you plan meals more easily. In short, a smart fridge saves time and reduces the mental load of keeping your kitchen organized by automating grocery management and meal planning.

Smart Coffee Makers: Your Personalized Morning Routine

A smart coffee maker is a must-have for coffee lovers looking to streamline their morning routine. These devices allow you to schedule when you want your coffee brewed, meaning it can be ready as soon as you wake up. Many smart coffee makers can be controlled through an app, allowing you to customize your brew from anywhere. You can adjust the strength and temperature, and even set it to brew at a specific time.

Some smart coffee makers integrate with voice assistants like Alexa or Google Assistant, so you can ask for a coffee to be brewed without getting out of bed. If you're running late, a smart coffee maker can have your coffee ready the moment you walk into the kitchen, reducing the time you spend on this daily task.

Using Smart Plugs with Your Non-Smart Appliances

Even if you don't have a fully smart-equipped home, you can still automate simple tasks with the help of smart plugs. These devices allow you to turn any standard appliance into a "smart" appliance. Simply plug the device into the smart plug, then connect the plug to your Wi-Fi network. You can now control the appliance remotely through an app or voice assistant.

For example, you can plug your slow cooker, toaster, or fan into a smart plug and schedule them to turn on or off at specific times. If you're a coffee lover who doesn't have a smart coffee maker, you can just plug a regular coffee maker into a smart plug. Then schedule it to turn on, or turn it on remotely, via an app. Of course, you do need to prepare the food or coffee for the machines to turn on, but automation still reduces effort and saves time.

Automate Home Maintenance by Scheduling Services

Maintaining a home can be a time-consuming and some-times overwhelming task. Sometimes it feels like there's always something that needs attention. However, automating your home maintenance gets essential tasks done on time without the mental load of constantly managing them.

Before we get into the details of how to automate your home maintenance, let's discuss why it is important to do so. Home maintenance is essential for preserving the value of your home, ensuring safety, and preventing costly repairs down the road. But it can often be overlooked or put off due to busy schedules. Automating your home maintenance tasks can ensure that they're done on time without needing to remind yourself or scrambling to make appointments. This automation can free up time and mental space for more important things, like spending quality time with your family or focusing on work.

There are several key areas of home maintenance that should be on your radar, many of which can be automated. In this section, we'll discuss how to streamline your home's maintenance through scheduling and automation, covering everything from HVAC tuneups to cleaning gutters. Here's an overview of the main categories:

- **HVAC system maintenance:** Your HVAC system needs regular maintenance to keep running efficiently and to prevent costly breakdowns. This includes changing filters, cleaning ducts, and scheduling annual tuneups for both your furnace and air conditioning. Automating HVAC maintenance ensures your home stays comfortable and energy-efficient year-round.

- **Plumbing:** Regular plumbing maintenance is vital to avoid leaks, clogs, or water damage. This might include drain cleaning, water heater checks, and inspecting pipes for signs of wear. Automating plumbing maintenance, like scheduling a yearly inspection or water heater flush, helps prevent big problems before they arise.

- **Roofing and gutters:** Cleaning your gutters and checking for roof damage are essential for preventing water damage, especially in regions with harsh weather. These tasks can be scheduled ahead of time to ensure that they're taken care of regularly, avoiding blockages and leaks that could cause long-term damage. I live in a very forested area and have to clean my roof and gutters two to three times a year.

- **Pest control:** Regular pest inspections and treatments can prevent infestations of insects or rodents. You can schedule seasonal visits to ensure that your home is protected year-round.

- **Appliance maintenance**: Major appliances like refrigerators, washers, dryers, and dishwashers require periodic cleaning and checks to ensure they run smoothly. Scheduling professional appliance maintenance, such as refrigerator coil cleaning or dryer vent cleaning, ensures these appliances stay in good working condition.

- **Lawn and landscape care**: Regular lawn care, including mowing, fertilizing, and irrigation system checks, is crucial for maintaining your outdoor space. Automated lawn care services can handle the task for you, keeping your yard neat and well-maintained without you having to lift a finger.

- **Safety equipment checks**: Regular maintenance of smoke detectors, carbon monoxide detectors, and fire extinguishers is critical for home safety. These are easy to overlook, but automating reminders to test, replace batteries, or schedule servicing can help ensure your home is safe and prepared in case of an emergency. I use lithium ion smoke/CO detectors that last 10 years and then need to be replaced. They warn you when they are about to expire. But I still put a reminder in my calendar for years down the road.

Scheduling Maintenance with Service Providers in Advance

One of the keys to automating your home's maintenance is scheduling services in advance. You can automate maintenance appointments in the following ways:

- **Set up recurring appointments:** For tasks like HVAC tuneups, pest control, or lawn care, most service providers will allow you to set up recurring appointments. For example, an HVAC company may offer an annual service plan where they automatically schedule your yearly inspection, filter change, and tuneup. With a recurring service, you don't have to worry about remembering when it's time for a check-up. Also, you can often negotiate a discount on service with recurring service agreements.

- **Use service provider apps:** Many service providers, from HVAC to pest control, have their own apps or platforms that let you book and schedule services. Some even allow you to set up automatic reminders for when it's time for your next appointment, helping you stay on top of maintenance.

- **Subscription services:** Certain home maintenance tasks, such as filter replacements or appliance cleaning, can be automated through subscription services. For example, companies like Filter Easy will send you new filters on a regular basis, so you don't have to remember when to change them yourself. Similarly, appliance maintenance services can be set up on a subscription basis, with professionals visiting on a set schedule for routine maintenance.

- **Smart home integrations:** Many smart home systems allow you to schedule reminders for home maintenance. You can integrate your home automation system with a calendar app or task management tool, sending alerts to your phone or voice assistant when it's time to schedule routine services like gutter cleaning, HVAC tuneups, or plumbing checks.

Leveraging Technology to Track and Manage Your Maintenance Tasks

In addition to scheduling and automating services, you can use technology to track the status of your home's maintenance needs. Here are some helpful tools and strategies:

- **Maintenance apps:** Apps like HomeZada and Househappy allow you to create a personalized maintenance schedule for your home. These apps can remind you when tasks need to be done, track service history, and keep records of repairs or improvements for future reference. You can also simply use your favorite general reminder app, like Todoist or Apple's Reminders apps.

- **Smart sensors and monitors:** Some smart home systems include sensors that can alert you when there's a potential issue with your home, such as a leaky pipe, high humidity levels, or a malfunctioning appliance. This can help you stay ahead of any potential problems and automate the response to these alerts, triggering maintenance services or sending reminders for you to take action.

- **Shared calendars and reminders:** If you live with others, a shared family calendar (like Google Calendar) can help keep everyone on the same page when it comes to home maintenance. You can set reminders for everyone to check things like air filters and smoke detectors, or schedule pest control services.

- **Smart appliances:** You can look for smart appliances that have servicing reminder apps. For example, my generator has an app that tells me when it's time for it to be serviced based on how much it's been used and when it was last serviced. My smart ecobee thermostat reminds me when it's time to change my HVAC's air filters and have my HVAC system serviced.

Schedule Your Car Maintenance

Cars, trucks, and motorcycles require regular maintenance to keep them working well and for longer. Combustion engine vehicles (we'll discuss electric vehicles later in this section) require oil changes, tire rotations, brake checks, and beyond; the list of maintenance requirements can seem endless. It's easy to miss maintenance without automating and scheduling it in some way.

Automating your vehicle's maintenance can save you time, reduce stress, and keep your car in good working condition, lessening the risk of breakdowns and expensive repairs. In this chapter, we'll explore the importance of regular vehicle maintenance, ways to automate it, and which maintenance tasks are most critical for vehicle longevity and safety. We'll also discuss how leasing a vehicle can simplify the maintenance process.

The Importance and Challenges of Vehicle Maintenance

Proper maintenance keeps your vehicle running smoothly and prevents problems from snowballing into costly repairs. Ignoring basic maintenance, such as changing the oil, can lead to engine damage, reduced fuel efficiency, and an increased risk of mechanical failure. Additionally, regular maintenance helps maintain the vehicle's resale value, ensures it passes inspection, and keeps you safe on the road.

The challenge, however, lies in the fact that vehicle maintenance is often irregular and complicated. If you've ever looked at the service schedules in the manual, you'll know that the vehicle needs different services at different times. Each car has its own maintenance schedule, with some tasks required more frequently than others.

For example, some vehicles require oil changes every 3,000 miles, while others may go 7,500 miles or more between changes. Similarly, tire rotations may be needed every 6,000 to 8,000 miles, and brake inspections are recommended every 12,000 miles. Many vehicles also require fluid checks (brake, transmission, power steering, coolant) at specific intervals, which may not be immediately obvious to the average car owner.

There are also tasks like timing-belt replacements or air filter changes that are necessary at certain intervals, often based on the vehicle's age or the manufacturer's recommendations. While it's important to follow these schedules to avoid costly repairs or breakdowns, keeping track of each item can become overwhelming, especially when they don't always align with regular intervals like oil changes.

The Most Important Services for Your Vehicle

While you should keep up on all the factory-recommended maintenance, there are a few services you should definitely not skip, as they are incredibly important to the functioning and longevity of your vehicle:

- **Oil changes**: Regular oil changes are crucial for maintaining your engine's performance. Oil lubricates the engine parts, reduces friction, and prevents overheating. If left unchanged for too long, the oil can become dirty and ineffective, potentially causing severe engine damage.

- **Brake inspections**: Your braking system is a very important safety component of your vehicle! Regular inspections will spot problems before they become serious.

- **Battery maintenance**: Your vehicle may not start if your battery is in poor condition, or if the connections to your battery are compromised. Some vehicles come with a system that alerts you when there are battery problems.

- **Fluid checks**: Your car relies on several types of fluids, including brake fluid, transmission fluid, coolant, and power steering fluid. These fluids need to be checked and replaced regularly to keep your vehicle in optimal working condition.

- **Tire maintenance**: Ensuring that your tires are properly inflated and rotated is critical for vehicle safety and performance. Regular tire rotations help ensure even wear, extending tire life and improving gas mileage.

You should try to automate these services as much as you can. Sure, other maintenance is important, but these are the services to prioritize.

How to Automate Your Vehicle Maintenance

Automating your vehicle's maintenance tasks is easier than ever, thanks to a variety of apps, reminders, and services designed to help you stay on top of your car's needs. Here's how you can automate your vehicle maintenance:

- **Use vehicle maintenance apps:** Many apps are available to help you track your vehicle's service needs automatically. Apps like Car Minder or AutoCare allow you to input your vehicle's information and keep track of all scheduled maintenance, including oil changes, tire rotations, brake inspections, and more. These apps send reminders when it's time for a service, based on either the mileage or time intervals you set. Some even integrate with your car's OBD-II port, which is typically located underneath the dash. It allows them to pull diagnostic data from your vehicle and alert you to any potential issues.

- **Set calendar reminders:** If you prefer a simpler approach, you can use a digital calendar (like Google Calendar or Apple Calendar) to schedule maintenance reminders. You can set recurring reminders for tasks like oil changes, tire rotations, and fluid checks. This ensures that you don't miss any important service dates, and it's easy to set reminders for every milestone your vehicle reaches.

- **Leverage dealership or service center reminders:** Many car dealerships and service centers offer maintenance-tracking as part of their service packages. Once your vehicle is registered with a dealership, they will send you automatic reminders about upcoming service appointments, warranty checks, and recalls. Many dealers also send out reminders for regular maintenance, like oil changes, based on your vehicle's mileage.

- **Subscription-based maintenance services:** Some companies offer subscription services for vehicle maintenance, taking care of all your scheduled maintenance at set intervals. These services often cover oil changes, tire rotations, fluid checks, and more, offering both convenience and cost savings. Maintenance subscription services reduce the need to track everything yourself while providing discounts for routine tasks.

You can still schedule maintenance, even if you do it yourself. Just take a look at the manufacturer's recommended scheduled maintenance schedule. Then put those maintenance tasks on your calendar, at least a year in advance. Use an online calendar that can send you reminders.

Leasing a Vehicle for Simplified Maintenance

Leasing a vehicle can be an effective way to simplify your vehicle maintenance responsibilities. When you lease a car, the maintenance tasks are often included in the lease agreement or covered under the vehicle's warranty, especially for newer models. Leasing companies also offer maintenance packages that cover routine services like oil changes, tire rotations, and even roadside assistance.

This means that with a lease, many of the regular maintenance tasks are scheduled and managed for you. In many cases, leasing companies may also handle recall notices and other urgent maintenance issues, giving you peace of mind that your vehicle is always taken care of. If you don't want to worry about long-term maintenance costs or keeping track of multiple service schedules, leasing can be a great option to simplify vehicle upkeep.

Electric Vehicle (EV) Maintenance

Many of the maintenance services discussed so far in this section are primarily for combustion engine cars. But what if you have an EV? One of the biggest advantages of owning an EV is that the maintenance requirements are typically much simpler, thanks to the absence of many of the moving parts found in a conventional engine. Unlike combustion engines, which rely on complex systems of belts, pistons, and fluids, EVs are powered by electric motors and batteries that require less frequent upkeep. Automating the maintenance of an EV, however, is still important to ensure the longevity of the vehicle, optimize performance, and keep everything running smoothly.

Key maintenance tasks for EVs are much less frequent than for gasoline-powered vehicles. For example, EVs don't need oil changes, timing-belt replacements, or exhaust system maintenance. However, some regular checks are still required, such as monitoring the health of the battery, maintaining tire pressure, and ensuring that the brake pads are functioning properly. Automating maintenance for an EV could involve setting up reminders for things like tire rotations or battery health checks. Many EVs offer integrated diagnostic tools that can send real-time updates about the battery's condition, tire health, and other essential systems. You can set these to automatically alert you when maintenance is needed or when it's time for a scheduled check-up.

Other Modes of Transportation

If you primarily get around by bicycle or scooter, these same concepts apply. A bike still needs regular tuneups (brake checks, gear adjustments, lubrication, wheel inspection, bolt tightening, cleaning, etc.). Many of these tasks you can learn to do yourself, or you can schedule regular appointments at your local bike shop. Two advantages of bikes, scooters, and the like is that maintenance is needed less often and is less expensive!

Of course, there will be times when you need to travel further afield. In the next chapter, we'll see how automation can streamline hotel bookings, flight check-ins, and car hire.

Key Takeaways

- **Smart home technology**: Automate everyday tasks at home, such as controlling the thermostat and managing lights and security systems, to save time and increase convenience.

- **HVAC and lighting:** Use smart thermostats and smart lighting to make your HVAC and lighting systems work for you with minimal intervention and maximum comfort.

- **Home maintenance automation:** Use apps or tools to schedule and track regular home maintenance tasks, such as cleaning, lawn care, and appliance upkeep.

- **Vehicle maintenance:** Schedule appointments and reminders in advance for important auto maintenance tasks to keep your vehicle in great condition.

CHAPTER 7

AUTOMATE YOUR TRAVEL

The journey of a thousand miles
begins with a single step.
—Lao Tzu

Traveling, especially air travel, can be a stressful and time-consuming experience. There are many steps involved that can drain your energy and eat up valuable time. Fortunately, there are numerous ways to streamline your journey so you arrive at your destination as smoothly and quickly as possible. In this chapter, we'll explore strategies for automating your air travel, including speeding up airport security, managing your frequent-flyer rewards, and using technology to handle logistics like parking and car services.

Speeding through Security: TSA PreCheck, Global Entry, and CLEAR

One of the most time-consuming and stressful parts of air travel is going through airport security. The good news is that several programs can help you speed up the process. The following will save you valuable time at the airport:

- **TSA PreCheck:** A trusted traveler program that allows you to bypass the long security lines at many airports. With TSA PreCheck, you don't have to remove your shoes, belt, or jacket, and you can keep your laptop and liquids in your bag. This expedited security process is available at most major airports in the U.S. and can be a huge time-saver, especially during peak travel periods. Once approved, your TSA PreCheck status can be automatically added to your airline profile, so it's applied to all your future flights.

- **Global Entry:** A must-have if you frequently travel internationally, it provides expedited entry into the U.S. at select airports and speeds up the customs process, allowing you to skip the long lines. Additionally, Global Entry includes TSA PreCheck benefits. It's a great way to minimize wait times when traveling abroad, and the application process is done online with an in-person interview.

- **CLEAR:** Another trusted traveler program that uses biometric data, like fingerprint and iris scans, to verify your identity and give you access to expedited security screening. CLEAR can also work in conjunction with TSA PreCheck, allowing you to bypass the initial ID check and get directly to the screening area. While TSA PreCheck focuses on providing fast screening through security, CLEAR speeds up the process of proving who you are. Additionally, CLEAR is used at major sports stadiums and arenas worldwide.

All these programs can be integrated with your travel profiles, so once you've enrolled, you don't need to manually enter your information every time you book a flight. Speaking of which...

Storing Your Frequent-Flyer Numbers and Rewards Programs

Frequent-flyer programs are one of the best ways to make the most of your air travel, but managing all your membership numbers, rewards, and loyalty points is cumbersome. You can automate this process by storing your frequent-flyer numbers and rewards program details in the online profiles you use to book flights. Here's how you can streamline the process:

- **Airline profiles:** Most airlines allow you to create an online account where you can store your frequent-flyer number and preferences. Add your frequent-flyer number to your profile, and it will be automatically applied when you book tickets. This means you don't have to remember to manually enter your number each time you book a flight.

- **Travel management apps:** Apps like TripIt or Award-Wallet help you organize your travel details, including frequent-flyer miles, hotel rewards, and car rental memberships. These apps can automatically track your points and send you reminders when you're about to reach a milestone, such as achieving elite status or needing to use expiring miles. Linking your travel rewards accounts to these apps simplifies the process and ensures you never miss out on valuable rewards.

- **Integrated booking services:** When booking flights through platforms like Google Flights, Expedia, or the airline's website, make sure your frequent flyer number is linked to your account or booking profile. This integration ensures that your rewards are applied automatically.

Scheduling Parking or Car Service

Another way to automate your travel experience is by handling parking and transportation needs in advance. Pre-scheduling these services will save you time and stress. Here are some ways to simplify travel to and from the airport:

- **Airport parking reservations:** Many airports offer online parking reservation systems where you can book a spot ahead of time. Services like ParkWhiz, SpotHero, and airport-specific apps allow you to reserve parking in advance. Some parking services even offer valet options or covered parking for additional convenience.

- **Car services and ride-sharing:** Automating your ride to and from the airport is another way to streamline your travel. Services like Uber, Lyft, and traditional car services allow you to schedule a pick-up in advance. Many of these apps let you choose specific pick-up times and set up recurring trips, which reduces hassle and stress. Additionally, some services like Uber or Lyft integrate with your calendar and can schedule your ride, based on your flight's departure or arrival time.

Automating Travel Logistics: Hotel and Car Rentals

Booking your accommodation and car rental can also be automated to save time. Ensure that you always have accommodations and transportation when you need them with the following:

- **Automated hotel reservations:** Many travel platforms, such as Booking.com, Hotels.com, and Airbnb, allow you to set up automatic booking confirmations, so you can reserve your hotel accommodations with just a few clicks. If you have loyalty memberships with hotel chains, linking those accounts to your booking platforms ensures that you get the benefits of rewards programs and any special offers without manually entering your details every time.

- **Automated car rentals:** When renting a car, the same automation process applies. You can link your frequent renter program (like Hertz Gold Plus Rewards or Enterprise Plus) to your car rental account, enabling potential upgrades or discounts with every rental. Some platforms even offer "one-click" reservations that allow you to book your car rental without needing to enter all of your details each time. Furthermore, some car rental programs allow you to bypass the check-in experience and just get your keys, go to your car, and drive away. This can be a true time saver!

Other Ways to Automate Travel

Beyond the main elements of booking flights, managing transportation, and earning rewards, there are several other ways you can automate your travel to make it even easier:

- **Travel notifications:** Set up automatic flight status alerts via your airline or travel apps. These notifications will keep you updated on any delays, gate changes, or cancellations, so you don't have to monitor your flight constantly. Apps like FlightAware or Google Flights can provide real-time updates and notify you when it's time to head to the gate.

- **Save your electronic boarding pass:** Before you travel, make sure to store your boarding pass in Google or Apple Wallet. This makes it convenient to pull up your boarding pass when needed. Unlike many airline apps, these services will also work when your phone doesn't have a stable internet connection.

- **Packing lists and travel reminders:** Apps like Pack-Point allow you to create custom packing lists based on your destination, weather, and travel plans. These apps can send reminders to pack specific items and help you avoid forgetting essential travel accessories. Or, you can just use something simple like Google Keep.

- **Custom travel routines:** Create automated travel routines with voice assistants. For example, you can set a reminder for when to check in for your flight, when to leave for the airport, and when to check the weather at your destination.

- **Travel agents for automated planning:** Many travel agents offer personalized planning services that can automate your entire trip—from booking flights and accommodations to creating itineraries and handling special requests. Travel agents can set up everything for you in advance, and they'll send you reminders and updates as your trip approaches, ensuring everything runs smoothly without the hassle of doing it all yourself. Some agents even offer concierge services that can adjust your plans as needed during your trip.

What I Do

I don't travel a lot, but I do travel enough, so making it efficient is more than worth it. Here are a few of the things I do to make travel less stressful and more time-efficient:

- **Travel sites:** I use a couple of online travel services and airline sites that store my known traveler number and my airline, car rental, and hotel reward account information. This ensures I get to use TSA precheck when available and that my rewards are properly credited whenever I book travel services.

- **CLEAR:** I also have a membership to CLEAR, which further streamlines my air travel experience. I've also used it to access shorter lines at sporting events.

- **Packing lists:** I don't need anything fancy here. I use Google Keep and just make simple packing checklists that I customize for each trip. This way, I don't forget to pack anything!

- **Work and credit card perks:** I take advantage of various travel perks I get through my employer and credit cards. They help pay for my Global Entry and CLEAR subscriptions, enrollment in premium car rental clubs, reduced parking rates, rideshare credits, and more.

- **Google Wallet**: I always save my boarding passes to Google Wallet, even if I can access it using the airline's app. Google Wallet will work even when I have no internet connection, and that has been handy a few times.

Traveling is often enjoyable, but keeping track of tickets, hotel reservations, and car bookings can be stressful and time-consuming. Automating these tasks not only saves time but ensures that nothing slips through the cracks at a crucial moment.

Booking travel arrangements often requires multiple online accounts. This can lead to problems if you forget passwords or your accounts are compromised—the subject of the next chapter.

Key Takeaways

- **Airport navigation:** Use services like TSA PreCheck and CLEAR to expedite security checks and reduce wait times, making airport experiences smoother and faster.

- **Make loyalty programs easy to use:** Store your frequent flyer numbers and rewards program details in the online profiles you use to book trips.

- **Car rentals and transportation:** Automate car rentals and transportation bookings through apps like Uber or Lyft, streamlining transportation logistics.

- **Travel reminders:** Set up automated alerts for check-in reminders, flight status updates, and other essential travel notifications to stay informed with minimal effort.

CHAPTER 8

AUTOMATE YOUR PASSWORDS FOR SECURITY AND CONVENIENCE

A strong password is the first line of defense against cyberattacks.
—Bruce Schneier

Managing passwords is both a necessity and a challenge because so much of your life is lived online. Each website, app, or other service requires a username and password, so in theory you need to remember countless credentials as well as make each unique and secure. Keeping track of all these passwords while ensuring they're strong enough to protect your personal information can quickly become overwhelming.

Fortunately, there are several ways to automate password management to enhance both security and convenience. In this chapter, we'll explore the importance of automating your passwords, the way to use password managers, and the way to set up two-factor authentication (2FA) to secure your online accounts effectively.

The Importance of Strong, Unique Passwords

Using strong, unique passwords for each of your accounts is one of the most important steps you can take to protect your personal data. Passwords like "123456" or "password" are far too common and are easily guessed by hackers. A good password should be at least 12 characters long. I recommend a phrase that's easy for you to remember but would be hard for others to guess, like "ISecretlyLoveTaylorSwift!".

One common mistake people make is reusing passwords across multiple sites. While this may seem like a time-saving tactic, it's actually a security risk. If one site is compromised, all accounts and other services using the same password are vulnerable. This is where password automation can help. Creating and storing unique passwords for each account secure each login without the burden of having to remember every password.

Using Password Managers to Automate Password Storage and Retrieval

A password manager is a tool that helps you store and organize all your passwords in a secure, encrypted database. The best part? Most password managers can generate strong, unique passwords for each of your accounts, saving you the trouble of coming up with them yourself. Password managers like 1Password, LastPass, and Bitwarden are great options for automating your password management. They can automatically fill in passwords for websites and apps, so you don't have to type them in every time.

By using a password manager, you need to remember only one master password (which should be strong, of course). The password manager then securely stores all your other credentials, allowing you to quickly and easily access your accounts. Many password managers offer browser extensions, mobile apps, and even integration with two-factor authentication (2FA) methods, which further automates the login process.

Creating Unique Passwords, Using an Algorithm

For those who prefer a bit more control or want to avoid using a password manager, an alternative is to create your own algorithm for generating unique passwords. This method allows you to come up with secure passwords on your own, while ensuring they remain distinct for each account.

One popular method involves combining a base word with the name of the website or service you're using. For example, you might choose a base phrase like "Secure1!@" and then append the first three letters of the website or service name. For a Netflix account, for instance, you could create a password like "Secure1!@Net". This way, each password is unique (because of the website-specific part) and complex, but you can easily remember it by following the same pattern.

The key to this method is ensuring that the base word or phrase you choose is strong and that the website-specific component is long enough to make the password unique. While this method provides a balance of security and convenience, it's still a bit manual compared to using a password manager. Still, it's an option for those who prefer more control over their passwords while ensuring they don't reuse them across sites.

Two-Factor Authentication (2FA): Adding an Extra Layer of Security

While strong passwords are essential, they're not enough on their own to guarantee your security. This is where multifactor authentication (MFA) comes in. Multifactor authentication generally describes security technologies and practices that require multiple methods of authentication in order to access a system. These authentication methods include:

- A password, passphrase, or PIN (personal identification number): *something you know*

- A token, smartcard, or key: *something you have*

- Biometrics like a fingerprint, voice recognition, or a retinal scan: *something you are*

You've probably been using multifactor authentication for years. When you go to an ATM machine to withdraw money, you insert an ATM card (something you have) and then enter a PIN (something you know). This is using multiple methods, or factors, to authenticate that you are you and can have access to your bank account.

Multifactor authentication can use two, three, four, or more methods of authentication. Two-factor authentication, or 2FA, is a type of multifactor authentication that requires only two methods. The ATM example above is 2FA. It's also 2FA when your bank emails or texts you a one-time authorization code to log in to their website.

There are several methods of 2FA, each with its own level of security and ease of use:

- **Text message (SMS) authentication:** With SMS-based 2FA, you receive a code sent to your phone via text message, which you enter along with your password. While convenient, SMS-based 2FA is considered less secure than other methods because text messages can be intercepted or your phone number can be compromised. This method is still better than using just a password, but it is not the most secure.

- **Authenticator apps (e.g., Google Authenticator, Authy)**: Authenticator apps generate time-sensitive codes that you use to log in after entering your password. These codes are not transmitted over the internet, making them more secure than SMS-based 2FA. You can set up these apps to automatically generate codes without requiring a network connection, which is particularly useful if you're concerned about security. Apps like Authy can even back up your 2FA settings, making them easier to recover if you lose access to your device.

- **Hardware keys (e.g., YubiKey)**: Hardware security keys, like YubiKey, are a physical device you plug into your computer or mobile device to authenticate your login. These keys offer a very high level of security because they cannot be easily hacked or phished. Once set up, hardware keys work seamlessly with compatible websites and services, and because they require physical access to the device, they're very secure.

- **Biometric authentication**: Some systems also support biometric authentication, such as fingerprints or facial recognition. These methods are increasingly popular for mobile apps and devices, providing an easy and secure way to verify your identity.

The most secure 2FA method is hardware keys (like Yu-biKey), followed closely by authenticator apps. While SMS is convenient, it's vulnerable to interception, so it should be avoided if possible. For automation, authenticator apps are the best option because they integrate seamlessly with your password manager or can be set up to automatically generate codes when needed.

Integrating 2FA with Password Managers

Most password managers support two-factor authentication, making the login process even more automated and secure. Once you've set up 2FA for your accounts, your password manager can store both your password and the associated 2FA codes (if you're using an authenticator app). This allows for smooth, automated logins. Some password managers even integrate with biometric authentication for an additional layer of ease and security. Using a password manager combined with 2FA allows you to secure all your online accounts without the hassle of remembering long, complex passwords or dealing with the inconvenience of multiple authentication steps.

What I Do

As a personal cybersecurity professional, I take the recommendations in this section to heart. I use a password manager to automate my passwords, since I have so many of them. Logging in is easy and secure.

I also use multifactor authentication whenever possible. I have two hardware security keys that I use whenever I can, but many of the services I use don't yet work with hardware security keys. In that case, I use 2FA authentication apps (specifically, Google Authenticator for personal accounts, and Microsoft Authenticator for work accounts). I'll use texts and emails for 2FA if neither security keys nor authentication apps are offered.

In the Real World: Automating Passwords for Security and Convenience

Jessica is a freelance writer, who manages multiple online accounts for her clients, social media, and personal use. She used to rely on simple passwords, often reusing them across various sites to make it easier to remember. But one day, she receives an email about a data breach from a site she frequents. Though she wasn't affected, the experience makes her realize just how vulnerable her online accounts are. Determined to enhance her security, Jessica decides to take control by automating her password management process.

First, she starts using a password manager (1Password). She installs the app on her devices and sets it up to generate a strong, unique password for each account she uses. Instead of remembering dozens of passwords, she needs to remember only one master password to access everything securely. With 1Password, Jessica automatically fills in her login information across websites, making it quick and easy to access her accounts without compromising security.

Next, she enables two-factor authentication (2FA) on all her important accounts—email, banking, social media, and work-related tools. For added convenience, she uses Authy to store her 2FA codes securely. Now, whenever Jessica logs in, she's prompted to enter a code from her Authy app in addition to her password, adding an extra layer of security that helps protect her from unauthorized access, even if someone manages to get hold of her login information.

Jessica has significantly reduced her risk of online breaches. The process of managing and securing her passwords is now automated and streamlined, allowing her to focus on her work without constantly worrying about password fatigue or security threats.

Automating the tedious task of setting and remembering passwords is a true time-saver—an absolute no-brainer solution for everyone, no matter how old you are or what you do. In the next chapter, we'll dive into how automation can benefit you at work.

Key Takeaways

- **Use a password manager:** Store and manage your passwords securely with tools like 1Password for easy access and to reduce the risk of forgetting or reusing weak passwords.

- **Strong, unique passwords:** Use a strong and unique password for each of your accounts, reducing the risk of weak or repeated passwords that can be exploited by attackers.

- **Two-factor authentication (2FA):** Enable 2FA for your accounts, using tools like Authy and Google Authenticator, to add an extra layer of security beyond just your password.

CHAPTER 9

MAXIMIZING WORK PRODUCTIVITY THROUGH AUTOMATION

Automation is the embodiment of
"work smarter, not harder."
—Allen F. Morgenstern

Every part of the job lifecycle, including looking for jobs, applying to them, and doing well in them can be tedious and filled with time-consuming tasks. You can automate each of these parts of your job to save time, be more efficient, and get more done no matter what your profession, whether you're a tradesperson or a white-collar office worker. In this chapter we'll cover automation techniques applicable to the entire job lifecycle.

Automating Your Job Search and Interview Prep

Job searching and preparing for interviews can be time-consuming and overwhelming, especially when you're juggling other responsibilities. Automation can help streamline both the job search process and your interview prep, allowing you to focus on what truly matters—finding the right job and performing your best in interviews.

First, let's discuss ways to automate your job search. The job search process often involves sifting through countless job boards, tracking applications, and setting up alerts for new job postings. Automating these tasks can save you time and keep you on top of opportunities without having to constantly check job boards. Here's how to do it:

- **Job search engines and alerts**: Use job search platforms like LinkedIn, Indeed, Glassdoor, and ZipRecruiter to automate your job search. These platforms allow you to create customized job alerts based on your preferred location, salary range, job title, and more. Set these alerts to notify you whenever a new job that matches your criteria is posted. You can even set frequency preferences (daily, weekly) so you don't get overwhelmed by notifications.

- **RSS feeds and aggregators**: Use tools like Feedly to subscribe to the RSS feeds of your favorite job boards or company career pages. This way, you can automatically see new job listings as soon as they are posted. RSS feeds can reduce the need to visit multiple job boards throughout the day, by aggregating all the company boards in one place.

- **AI-powered job matching**: Some job platforms, such as LHH Recruitment Solutions or Jobscan, use AI to match your profile with job opportunities. After you upload your résumé or fill out your professional information, these platforms suggest jobs tailored to your skills, experience, and preferences. This means that the system is actively working on your behalf and you don't have to manually search and sort through as many irrelevant listings.

- **Automated résumé and cover letter tailoring:** Tools like Jobscan also help automate the process of tailoring your résumé and cover letter to each specific job posting. These tools analyze the job description and compare it to your résumé and then offer suggestions to optimize your documents for each application. Submitting a customized résumé increases your chances of passing through applicant tracking systems (ATS) and catching the attention of hiring managers.

Networking is a critical part of the job search process, but it is often a time-consuming activity. Automating aspects of your networking can help you stay connected and maintain meaningful relationships without the constant pressure of actively reaching out. Although I have the skills to do a good job networking on my own, I often lack the drive and motivation to do it. These tools and techniques can be a tremendous support for your networking efforts:

- **LinkedIn networking:** Use tools like LinkedIn Helper or Dux-Soup to automate the process of connecting with potential employers, recruiters, and industry professionals. These tools can help you send personalized connection requests, follow up on conversations, and engage with content—all while saving time. I've gotten a couple of jobs in the past through networking and searching on LinkedIn.

- **Email automation for outreach:** Automate your outreach emails with tools like Mailshake or HubSpot. These platforms allow you to send personalized follow-up messages, thank-you emails, or check-ins with just a few clicks. For example, you can automatically send a thank-you email after an interview or follow up with hiring managers if you haven't received a response in a few weeks. You can also track your emails, including whether or not they were opened, and reuse messages (or parts of them) in other emails.

After researching jobs and applying for them, you'll be asked to do interviews. Interview preparation is important for making a strong and positive impression that will increase your chances of receiving a job offer. Automation can help you prepare more efficiently and consistently for interviews, so you can focus on delivering your best performance. Here are some job interview automation tips:

- **Interview question practice**: Use AI-powered interview prep tools like Interviewing.io or Pramp to simulate mock interviews. These platforms offer automated practice sessions where you can answer common interview questions in real time. You'll receive feedback from AI or real interviewers, helping you refine your responses and improve your delivery.

- **Automated skill assessments**: Platforms like Codility, HackerRank, or LeetCode provide coding assessments and tests for technical interviews, while sites like Vervoe offer skill assessments for various industries. These platforms automate the process of testing your skills, allowing you to practice and demonstrate your abilities before an interview, potentially increasing your confidence.

- **Preparation reminders**: Set up automated reminders in your calendar or task management apps like Google Calendar or Todoist to prepare for interviews.

These reminders can include tasks like researching the company, reviewing the job description, practicing answers to common questions, and preparing thoughtful questions to ask the interviewer. You can also schedule specific time blocks for different aspects of prep, such as a 30-minute session to research the company and a 20-minute session for practicing your elevator pitch.

- **Automated document storage:** Store important interview documents such as your résumé, cover letter, references, and portfolio in cloud storage services like Google Drive or Dropbox. With everything easily accessible, you can quickly retrieve the right documents whenever an opportunity arises. Many services offer automatic backups, ensuring you never lose your interview materials.

- **AI-generated feedback**: After practicing with automated interview prep tools or mock interviews, use platforms like Big Interview that offer AI-generated feedback on your answers. These tools analyze your responses, body language, and tone to provide insights into how you can improve.

One of the most helpful ways to automate your job search and interview prep is by staying organized. Tools like Trello or Airtable can help you track your applications, interviews, and follow-ups. You can set up boards or tables to track whom you've applied to, the status of your application, and when you need to follow up. Automating this ensures that no job opportunity slips through the cracks and you'll always know what to do next.

The tools out there to help you search for a job and perform better in interviews can give you amazing results. I recommend getting familiar with them before you need them for a job search. The stress of learning the tools when you are in need of a job can make you less productive.

In the Real World: Automating Job Search and Career Development

Nina has been considering a career change but finds the job search process daunting and time-consuming. She decides to automate much of it with LinkedIn Careers and Jobscan, which helps her tailor her résumé to specific job descriptions. She sets up LinkedIn job alerts for roles in her desired new field, and Jobscan helps her optimize her résumé for each position. Every time she applies, Jobscan analyzes the job description and recommends changes to improve her résumé's chances of passing through applicant tracking systems. Nina also automates follow-ups and interview reminders, using the Google Calendar integration with LinkedIn. Within a few weeks, Nina secures interviews at multiple companies and lands a job offer. Nina was able to use automation to focus on preparing for interviews instead of spending endless hours sifting through job listings.

Using Automation Techniques and Tools to Improve Your Work Productivity

Looking for and securing a job are the first steps in the process and only the beginning of how automation and artificial intelligence can save you time and improve your work performance. Earlier in this book, I talk about forming automatable habits. The workplace is a great opportunity to practice and perfect those skills. In this section, we'll discuss some of the best automatable habits to boost your work productivity.

Blocking Time for Specific Tasks

One of the most effective work productivity habits you can automate is time-blocking. Time-blocking involves scheduling specific chunks of time for particular tasks and ensuring you have uninterrupted time to focus on high-priority activities. Time-blocking helps you avoid multitasking and manage your day more efficiently.

Here are a few ways to automate the habit of time-blocking:

- **Use calendar apps:** Tools like Google Calendar or Outlook allow you to block off time for specific tasks. For example, you could schedule a 9:00 a.m.–11:00 a.m. block for focused work on a report or a 1:00 p.m.–2:00 p.m. slot for email management. The key is to treat each task as an event, which helps you maintain structure and prioritize your most important activities.

- **Set recurring blocks:** Automate your recurring tasks by setting up repeating time blocks. For example, every Friday afternoon, block time for weekly planning or catch-up on emails. This helps you stay on top of regular tasks, and they don't slip through the cracks as a result.

Creating and automating time blocks reduces decision fatigue and ensures that your most important tasks are given the attention they need.

Setting "Do Not Disturb" Hours on Your Phone and Apps

One of the biggest productivity killers is constant interruptions. This is a byproduct of the "attention competition" problem I detailed earlier in this book. Notifications from your phone, email, and apps can pull your focus away from critical tasks, reducing both the quality and speed of your work. Automating "Do Not Disturb" hours is a great way to carve out uninterrupted focus time. Here are a couple of tips for reducing distractions from your phone and apps:

- **"Do Not Disturb" on phones:** Most smartphones, including iPhones and Android devices, have a "Do Not Disturb" feature that can be scheduled automatically. Set your phone to go into "Do Not Disturb" mode during specific times, such as during deep work sessions or meetings. You can customize the feature to allow urgent calls or messages through.

- **Mute notifications on apps:** Apps like Slack, Teams, and email clients allow you to mute notifications or set "away" statuses. This stops you from being constantly distracted by incoming messages. You can automate these settings, based on your calendar events, or set custom periods for when you're focusing on specific tasks.

Automating your "Do Not Disturb" hours allows you to take control of your focus and reduce the stress of managing constant interruptions.

Scheduling Days to Work Remotely for Deep-Thinking Work

Some tasks, such as strategic planning, problem-solving, or creative brainstorming, require undisturbed, deep-thinking time. Working remotely or in a quiet, less distracting environment can be a great way to foster the focus needed for these types of work. If you regularly have the opportunity to work remotely in your occupation, you should take advantage of it. Here are some automation tips and other advice to help you make the most of remote working:

- **Automate remote work days:** Use your calendar or scheduling tool to set specific days for remote work, such as every Wednesday or the first Monday of the month. Working from home or a quiet place may give you more control over your environment, allowing you to eliminate distractions and to focus on deep-thinking tasks without the interruptions of office noise or meetings.

- **Set clear boundaries:** When working remotely, set clear expectations with your team or manager about your availability. Automate status updates on apps like Slack or Teams to reflect when you are in "deep work" mode. This communicates your schedule to others and ensures they understand when you're focusing on tasks that require minimal distractions.

Automating your remote work days helps you create a predictable routine for completing high-level tasks. This will help with the quality of your work and make you more productive.

Automating Daily and Weekly Review Processes

Productivity is not just about managing tasks throughout the day; it's also about taking time to reflect and adjust. A regular review process can help you evaluate your progress, prioritize upcoming tasks, and keep you aligned with long-term goals. Here are some automatable review habits you should consider:

- **Daily review:** At the end of each day, automate a 10-minute review session. Set a recurring task in your task management app or a recurring appointment in your calendar to remind you to review your completed tasks, assess what worked, and plan for tomorrow. This allows you to close out your day productively, leaving you prepared for the next day.

- **Weekly review:** Similarly, automate a weekly planning session to evaluate your goals, check in on ongoing projects, and adjust your priorities for the upcoming week. Building this habit helps you stay on top of your work and make progress toward your goals.

A regular review process makes you proactive instead of reactive, and automating it ensures you stick to the habit without needing to remember it each time.

Automating Email Management

Email can be one of the most time-consuming and costly tasks, with constant incoming messages, notifications, and the presumed need for quick responses. Automating your email management can help you stay organized and reduce the mental load associated with managing a constantly filling inbox. Here are some techniques that you can use regularly and automate to make email more manageable and take up less of your time:

- **Email filters and folders:** Use your email provider's filtering tools to automatically sort and categorize incoming emails. For example, you can set rules that move newsletters or less urgent emails into specific folders, while prioritizing important emails from clients or managers to stay on top of critical messages.

- **Pre-scheduled responses:** If you find yourself sending the same responses repeatedly, tools like Gmail's Canned Responses or Outlook's Quick Steps allow you to pre-write and automate common replies.

Automating your email management helps you stay organized, prioritize tasks, and reduce time spent sifting through unnecessary messages.

Managing Workplace Messaging Tools

Messaging tools like Slack, Teams, and other workplace communication platforms have become essential for collaboration, but they can also be major sources of distraction if not managed properly. It's important to learn good techniques for automating and managing these tools to maintain your productivity, and sanity! Constant notifications, group chats, and threads can pull your focus away from important tasks, leading to reduced productivity. Fortunately, automation offers several strategies to manage these tools more effectively.

Customizing Notifications for Priorities

One of the most effective ways to reduce distractions from messaging tools is by customizing your notification settings. Most communication platforms, including Slack and Teams, allow you to control how and when you're alerted to new messages. You can do the following:

- **Set notification preferences**: Both Slack and Teams let you control which messages send notifications and how they are delivered. You can set notifications to alert you for mentions only (e.g., when someone tags you with "@yourname") or for direct messages.

- **Mute non-essential channels**: In Slack and Teams, you can mute notifications from specific channels or conversations that are not directly related to your work. Muting less important channels reduces unnecessary noise and frees up mental space to focus on the channels that matter.

- **"Do Not Disturb" (DND) mode**: Both platforms offer a "Do Not Disturb" feature that lets you turn off all notifications for a set period.

Automating notification settings ensures that only critical alerts grab your attention, allowing you to focus on your work without interruptions.

Setting Clear Boundaries with Status Updates

Another way to reduce interruptions from Slack, Teams, or any messaging tool is by using the status update feature. This simple habit can automate your availability and let your colleagues know when you're deep in work or unavailable. You can do the following:

- **Set your status automatically**: Both Slack and Teams allow you to set a custom status that communicates your current work mode. For example, you can set your status to "In a Meeting," "Do Not Disturb," or "Focusing on a Project." This helps others understand when you're available and when you prefer not to be interrupted. You can also set an automatic status update, based on your calendar events, so it updates automatically when you're in a meeting or blocked off for focused work.

- **Scheduled status changes**: Some messaging platforms, like Slack, offer a feature where your status can be scheduled to change automatically. For example, you can set a "Focus Time" status during your dedicated work hours or change it based on your calendar to reflect your availability. This reduces the need to manually adjust your status and clarifies your work boundaries.

Automating status updates creates a barrier that helps protect your focus.

Automating Responses for Routine Interactions

Another way to reduce distractions is by automating your responses to frequently asked questions or routine interactions. This saves time and reduces time spent on replying to similar messages. You can do the following:

- **Use saved replies**: In Slack, create saved replies for frequently asked questions or common messages. For instance, if you often get asked for project updates or meeting links, you can set up a few pre-written responses that can be quickly sent with a click. This way, you don't have to spend time crafting individual responses every time someone asks.

- **Use bots for routine tasks:** Both Slack and Teams allow you to integrate bots that can automate common tasks. For example, you can set up a bot that answers basic questions, like office hours or meeting schedules, so you don't have to respond directly.

You can reduce the number of distractions from messaging platforms by automating routine interactions. This will give you more time to focus on meaningful tasks.

Group Conversations and Thread Management

While team-wide chats can be essential for collaboration, they can also become overwhelming and noisy. Managing group conversations and threads effectively is key to reducing distractions. Doing the following helps:

- **Encourage threads:** In both Slack and Teams, you have the option to reply in threads instead of responding directly in the main channel. Encouraging your team to use threads for specific discussions can help keep the main conversation area less cluttered and make it easier to follow important messages. Set up automated notifications to alert you when someone replies to a thread you're following. This ensures you don't miss important updates while keeping the main chat area clean.

- **Organize channels:** Set up separate channels for different projects or teams in Slack or Teams. This allows you to mute or limit notifications from non-essential channels, reducing distractions from irrelevant conversations. You can also create a separate channel specifically for announcements so that important updates don't get buried in day-to-day chatter.

Organizing conversations and automating where you receive updates helps you stay focused on the tasks at hand.

Time-Boxing Communication

Time-boxing is a strategy that involves setting aside specific blocks of time during the day for checking and responding to messages. This prevents you from constantly checking messages throughout the day and getting pulled out of your focused work.

- **Dedicated communication time:** You can schedule designated times in your calendar, such as 30 minutes at the start and end of each workday, to go through Slack or Teams messages. By automating this habit and ensuring that you check messages at these set times only, you reduce the urge to constantly monitor your phone or app for new messages.

- **Automate follow-ups:** Tools like Zapier can help you automate follow-up reminders for messages you've sent but haven't received a response to. You can set up a rule that automatically reminds you to follow up with a colleague if they haven't replied within a certain time frame.

Time-boxing your communication reduces interruptions, allowing you to dedicate focused time to other important tasks without the anxiety of unread messages piling up. This is a good strategy to use with your email inbox as well.

Scheduling Breaks and Downtime

Rest is important to doing good work, but it's easy to forget to take breaks during a busy day. Scheduling time for relaxation or a walk can help refresh your mind and improve productivity when you return to your tasks. Go ahead and do the following:

- **Automate break reminders:** Use tools like Pomodoro timers or task management apps to schedule short breaks during your workday. You can also put them as recurring appointments in your work calendar. For example, you might work for 25 minutes and then take a five-minute break and schedule a couple of 15-minute stretching or walking breaks each day. Automation tools like Toggl or Focus Booster can track this for you, so you don't have to remember to stop working.

- **Schedule downtime:** Plan to stop after a certain number of hours of work. Automate reminders to take an extended lunch or leave work at a specific time to ensure you prioritize your health and well-being. You should put these in your work calendar as well.

Automating your breaks and downtime can reduce burnout and maintain high levels of focus when you are working.

What I Do

At work, I can easily get bogged down in meetings if I don't reserve time in my calendar to get other things done. I have regular time blocks in my work calendar for lunch, priority items, and prepping for the day.

I'm a big believer in taming the beast that is work email. I have filters set up to send various emails to different folders and then I set times for when I will look at them. For instance, I get alerts from various systems around work; depending on the severity of them, I have them go to an alert folder. Another example is that I receive lots of bills and invoices that are usually not urgent, so those go to their own folder.

I also keep my phone on "Do Not Disturb" during business hours. However, messages from contacts like my children and wife will give me a notification, so I won't miss anything urgent.

Work unavoidably consumes a large amount of your time. And while a life of leisure may be a distant dream, automation can at least make your office hours less stressful. There are plenty of time-consuming tasks to be done when you're not at work—eating, shopping, and caring for people and pets. We'll cover how to do these tasks more efficiently in the next chapter.

Key Takeaways

- **Automate time management**: Use scheduling tools like Google Calendar or Outlook to automatically manage meetings, block out focused work time, and help you dedicate time to the right tasks.

- **Implement regular work reviews**: Use planning sessions and reviews to proactively prioritize your work instead of being reactive to lower-priority items.

- **Email automation**: Use filters, templates, and autoresponders in your email client to automatically sort and respond to emails, reducing the time spent managing your inbox.

- **Manage workplace messaging**: Customize notifications and your work status to minimize distractions so you can focus on the work you need to complete.

CHAPTER 10

EXPLORING OTHER OPPORTUNITIES FOR AUTOMATION

*Automation is not just a technology,
but a mindset that embraces change and innovation.*
—Bob Crissman

I've covered a lot of areas in life that you can automate to give you more time for things you prefer to spend time on, but there are many more. You truly are limited by your imagination only. As technology continues to improve, more and more things will be easily automatable. Here are a few other life automation areas that you should consider.

Automate Your Shopping

Do you love shopping, or do you view it as a necessary evil? Perhaps, like me, you are somewhere in between. No matter where you are on the spectrum, you can save time and money (which can allow you to shop more if you'd like) by using various automations, services, and techniques. The rise and prominence of online shopping in the last couple of decades has made automating your shopping easier than ever.

Automate Purchases of Regularly Used and Consumed Items

A lot of the items we shop for are things we regularly use and buy. Essentials like personal care items, groceries, and various household supplies are bought regularly, and usually without a lot of variance. These are the types of purchases that are easy to automate.

One of the easiest ways to automate regular purchases is through subscription services. Many retailers and online stores offer automatic delivery options for products you buy frequently. This means you can set it up once, and products will arrive at your door on a schedule that works for you—be it monthly, quarterly, or even bi-weekly. Here are some subscription services that you can use for general shopping—everything from toiletries to medical, health, and wellness products:

- **Amazon's Subscribe & Save:** Amazon offers a Subscribe & Save option, which allows you to set up recurring deliveries for household products, toiletries, and even health and wellness items. You can choose the frequency of delivery, and Amazon will automatically send your order, often at a discounted price. This is perfect for items like paper towels, laundry detergent, vitamins, and toothpaste. Once you set up your order, you don't need to worry about remembering to reorder—everything is taken care of. I use this service for protein bars, loofahs, cleaning supplies, toilet paper, vitamins, and many other items.

- **Target's subscription service:** Target offers subscriptions for regular purchases. Its Target Subscriptions program offers a wide range of items, from cleaning supplies to baby products, which are automatically shipped on a schedule that suits you. Target even allows you to adjust delivery times and quantities as needed.

- **Beauty and personal care subscriptions:** Services like Birchbox or Ipsy provide personalized beauty and grooming products delivered on a regular basis. You can automate the delivery of skincare items, makeup, and personal care products like razors and deodorants, based on your preferences. This is especially helpful for busy individuals who want to keep their grooming supplies stocked up.

- **Walmart:** Walmart's app allows you to create recurring orders for frequently purchased items. Once you've added products to your cart, you can choose to set up recurring deliveries, making it easy to keep stock on essentials. Walmart's automatic reorder feature is perfect for things like pet supplies, cleaning products, and office supplies.

- **Chewy:** For pet owners, Chewy is a great option for automating regular pet product purchases, from food and treats to toys and grooming supplies. You can set up scheduled deliveries for your pet's favorite food or other essentials. You can also adjust the frequency of your deliveries, depending on how much your pet goes through each month.

- **Home Depot and Lowe's:** If you frequently buy tools, home improvement supplies, or garden products, both Home Depot and Lowe's offer automated delivery services for regular purchases. Their apps let you schedule deliveries for things like light bulbs, batteries, or even seasonal gardening items, so you always have what you need for your DIY projects or home maintenance tasks.

- **Health supplement subscriptions**: Services like Vitamin Shoppe allow you to set up regular deliveries for vitamins and supplements. These services provide personalized supplement plans based on your health needs and can automatically send you the products you need each month, saving you the hassle of reordering.

- **Automated prescription refills**: Many pharmacies, like CVS and Walgreens, offer automatic prescription refills. You can set up an account that automatically fills your prescriptions and even schedules deliveries. This ensures that you never run out of important medications or health supplies, giving you peace of mind and helping you stay on track with your health regimen. Depending on your insurance carrier, you may also have access to mail order prescription fulfillment.

Grocery Shopping and Meal Planning

Grocery shopping and meal planning can be some of the most time-consuming tasks in managing a household. Frequently creating shopping lists and figuring out what to cook for the week quickly adds up to a lot of time. However, by automating your grocery shopping and meal planning, you can save time, reduce stress, and ensure that you're always prepared for the week ahead.

Automating Email-Planning for Grocery Lists

Keeping track of your grocery needs can also be made easier by automating email reminders. Services like Amazon's Subscribe & Save or Instacart's shopping list allow you to receive email reminders for recurring grocery orders. You can automatically renew household staples like toilet paper, coffee, or snacks, without needing to manually reorder them each time. These reminders make it easy to stay on top of what you need and to avoid running out of essential items.

Additionally, grocery services and apps like Google Keep or Evernote can automate the creation of grocery lists by allowing you to quickly capture items you're running low on via voice commands or quick notes. These lists can be synced across devices, ensuring you always have an up-to-date grocery list at your fingertips, ready for your next shopping trip or delivery order.

Automating Grocery Shopping

Say goodbye to the frustration of trying to remember your shopping needs at the grocery store. Automation ensures you keep your refrigerator and pantry stocked with significantly less effort. Here are some tips to make this happen:

- **Use grocery delivery services:** Services like Instacart, Amazon Fresh, and Walmart Grocery allow you to automate your grocery shopping by creating recurring shopping lists based on your preferences or past purchases. These platforms let you order groceries online and have them delivered to your doorstep, saving you time spent at the store. You can set up weekly or bi-weekly deliveries for staples like milk, bread, and eggs, and make quick adjustments as your needs change.

- **Use smart shopping lists:** Many grocery store apps or meal-planning services, such as AnyList or Out of Milk, let you create and store shopping lists that are accessible on your phone. These apps automatically update your list as you run out of items or add new ones, and some can even sync with family members or housemates, ensuring everyone is on the same page when it comes to what needs to be purchased. You can also do this with voice assistants like Amazon Alexa and Google Assistant.

- **Place recurring grocery orders:** Some grocery delivery services let you set up automatic recurring orders for common items. For instance, if you frequently buy certain brands of pasta or snacks, you can set these items to be reordered at regular intervals. This eliminates the need to re-enter items you already buy, saving you time by not having to manually reorder each item every week.

- **Automate price-tracking and comparison:** Apps like Flipp allow you to track grocery prices at local stores and automatically compare prices across various retailers. These apps can alert you when an item you regularly purchase goes on sale, helping you save money by getting the best deal.

Automating Meal Planning

Meal planning is another area where automation can save you hours each week. Apps and services that create meal plans and shopping lists help streamline your cooking processes, make planning simpler, and reduce your prep time in the kitchen. Here's how these apps and other services can simplify meal planning and preparation:

- **Plan meals:** Services like Plan to Eat and Mealime allow you to plan your meals for the week, generate shopping lists based on your menu, and suggest recipes based on what you already have in your pantry. You can customize your meals according to dietary preferences (vegetarian, gluten-free, etc.), and the meal plans update automatically based on your input. With these tools, meal planning becomes as simple as picking out a few recipes.

- **Get ingredients delivered:** If you prefer to be more hands-off with meal planning, consider subscribing to a meal kit service like HelloFresh, Blue Apron, or Home Chef. These services deliver pre-portioned ingredients and recipes straight to your door, saving you the trouble of planning meals and shopping for ingredients. They also help you avoid food waste by providing exactly what you need for each meal.

- **Prepare meals, using smart devices:** Purchase smart kitchen devices that save time and reduce the hassle of cooking. Smart slow cookers like the Instant Pot or Crock-Pot can be programmed to cook meals while you're away or working. Similarly, smart ovens, like the June Oven, allow you to set cooking programs or even monitor cooking remotely from your phone.

- **Integrate grocery lists with meal plans:** Apps like Paprika and Cookpad integrate directly with your meal planning, allowing you to generate grocery lists automatically based on the recipes you choose. These apps let you store favorite recipes, organize them into categories (e.g., quick meals, vegetarian, etc.), and quickly create shopping lists for your planned meals.

Automating Grocery and Meal Planning Together

You can save time by automating aspects of your meal planning and grocery shopping separately. However, the real power of automation comes when you combine both your grocery shopping and meal planning into one seamless system. Many apps and services integrate shopping and meal planning into a single platform that's easy to use. They make it easy to set up the following:

- **End-to-end automation**: Apps like Eat This Much offer both meal planning and grocery shopping integration. They allow you to set your dietary preferences and generate weekly meal plans. Then, they automatically create a grocery list that can be ordered through delivery services or printed for in-store shopping. This end-to-end solution saves you time and energy, from deciding what's for dinner to making sure you have everything you need to cook it.

- **Synchronized services**: For more customization, sync grocery apps with meal-planning services. For example, syncing Mealime with Instacart allows you to plan meals and automatically send your shopping list to the delivery service. This way, you're not only automating your meal planning but also ensuring you get ingredients delivered to your door without additional effort.

In the Real World: Automating Your Meal Planning and Grocery Shopping

Eric is overwhelmed by meal planning and grocery shopping. He wants to eat healthier but finds it time consuming to figure out what to cook every week. He subscribes to HelloFresh, which automatically generates meal plans and sends him the ingredients needed to cook fresh meals each week. Eric syncs his HelloFresh plan with his Instacart account, which automatically orders the ingredients for delivery on a set day each week. Now, instead of spending hours figuring out what to make and shopping for groceries, Eric receives fresh ingredients, pre-portioned for the week, and doesn't have to set foot in a store. Eric enjoys home-cooked meals and spends less time on mundane food prep tasks by automating his meal planning and shopping.

Maintain Your Important Relationships, Using Automation

Important relationships in your life are worth putting effort into. Relationships need attention and work to flourish, but the good relationships are more than worth it. Automation can help build and nurture bonds with your friends and families.

One of the most effective ways to maintain relationships is by scheduling regular meetings in advance. Life can get busy, and it's easy for good intentions to get lost in the shuffle. I'm guilty of not staying in timely communication with loved ones. Scheduling recurring phone calls, video chats, or even in-person meetups ensures that you make time for the people who matter most. For example, you could set up a monthly lunch date with a close friend or schedule a weekly check-in with family members. These recurring appointments can be automated on your calendar, so you don't need to rely on memory. With tools like Google Calendar or Outlook, you can set up recurring events that automatically remind you to connect with others, even if it's just for a quick chat.

Additionally, automation can be incredibly helpful for remembering special occasions such as birthdays, anniversaries, or other milestones. You don't want to be the spouse that forgets your anniversary or your significant other's birthday. With the help of digital calendars, you can set reminders well in advance so you're never caught off guard when it's time to send a card, gift, or a thoughtful message.

Most calendar apps have notifications, which you can set for birthdays, anniversaries, and the like, giving you time to prepare. Receiving automated reminders eliminates the stress of scrambling to find the perfect gift at the last minute or forgetting an important milestone. You can even use services like Amazon or Giftagram to automate gift-giving by scheduling deliveries or selecting gift options for your loved ones in advance. These services help you choose personalized gifts that are sent directly to the recipient, saving you time while still making the gesture feel meaningful.

Another way to keep important relationships strong is by automating thoughtful reminders to reach out. Sending a text to check on a friend, calling a family member just to say hello, or scheduling regular holiday greetings are meaningful gestures for the important people in your life. Automated reminders keep you connected without needing to plan everything in advance. Setting reminders in your phone, or using apps like Todoist and Evernote, gives you plenty of opportunities to show appreciation, even when you're busy with other priorities.

Automate Your Pet Care

Taking care of pets can be incredibly rewarding, but it also requires a significant amount of time and effort. You have to feed them, ensure they get enough exercise, and schedule regular vet visits. No wonder pet care feels like a never-ending task. Automating certain aspects of your pet's routine can save time and reduce stress while making sure your pet receives the care they need, even when you're busy or away.

Automating Pet Feeding

One of the most time-consuming pet-care tasks is feeding pets at the right times. Luckily, automatic pet feeders are a great way to streamline this process. Smart pet feeders, such as PetSafe, WOPET, and Petcube Bites, allow you to schedule meals for your pet and control feeding times from an app on your phone. These feeders dispense the right portion of food at the appropriate time, helping to regulate your pet's eating habits and ensure they receive their meals on schedule, even when you're not home.

For multi-pet households, some feeders offer options for portion control. You can customize the amount of food each pet gets, which is particularly useful if they have different dietary needs. Some of these devices have built-in cameras so you can check in and ensure your pet is eating, making it easier to monitor their behavior from anywhere.

Automating Pet Exercise and Playtime

Keeping your pet active is essential for their health, but daily walks or play sessions can be challenging with a busy schedule. Fortunately, technology has made it easier to automate pet playtime and exercise. Even if you enjoy play and exercise with your furry little loved one, these tools and techniques can take the stress out of doing it when you aren't available:

- **Automated pet toys:** Devices like the PetSafe Automatic Ball Launcher or the iFetch allow your dog to play fetch on their own. These devices launch a ball for your pet to retrieve, providing mental and physical stimulation. These automated toys can be set to specific intervals so your pet gets regular playtime while you're at work or occupied with other tasks.

- **Smart cameras:** Internet-connected cameras like the Furbo Dog Camera or Petcube Play allow you to monitor your pet's behavior and ensure they're getting the necessary attention. Use them to remotely play with pets, communicate with them, and even reward them using built-in treat dispensers.

- **Automated walkers:** For smaller pets, devices like dog treadmills can help them get their daily exercise. These treadmills are designed for pets to walk or run safely indoors, making it easier to maintain their fitness routine when you don't have time to make it outside.

Automating Pet Health and Wellness

Taking care of your pet's health is an ongoing task that includes regular check-ups, grooming, and medication. Automation can make managing these tasks easier and more consistent. Some of the concepts mentioned below are very similar to the health and wellness advice that I wrote about earlier in this book. Available gadgets include the following:

- **Automatic litter boxes:** If you have a cat, a device like the Litter-Robot does away with constant manual scooping. These self-cleaning devices automatically scoop waste after each use. Many models are equipped with sensors to ensure that the litter stays fresh for your cat, maintaining a clean, odor-free environment with minimal effort on your part.

- **Health-monitoring devices:** Some pet care devices track your pet's wellness. Collars like FitBark or Whistle Go monitor your pet's activity levels, sleep patterns, and even their diet, providing insights into their overall health. These devices sync with your phone, allowing you to keep an eye on your pet's well-being, set goals for activity, and receive alerts if something seems off.

- **Automated grooming:** While grooming can be a regular, hands-on task, you can automate it to some degree with tools like self-cleaning brushes or pet-grooming robots. Devices like the PetSafe Self-Cleaning Pet Grooming Tool help remove loose hair and dirt without the mess. Some robotic grooming devices even allow you to schedule brushing sessions, ensuring your pet's coat stays clean and healthy.

Automating Veterinary Visits and Appointments

Scheduling and keeping track of veterinary appointments can be overwhelming, especially if you have multiple pets. Automating these appointments allows each of your pets to get the necessary medical attention without having to remember each visit. Instead, you could do the following:

- **Automate appointment reminders:** Many veterinary clinics offer email or text reminders for vaccinations, check-ups, or treatments. You can set up automatic notifications to ensure you never miss an important appointment. Some apps provide easy access to scheduling and reminders directly from your phone, streamlining the process of keeping track of your pet's health needs.

- **Automate pet insurance:** Use apps from insurers like Trupanion or Petplan to automate claims and track your pet's medical records. They allow you to submit medical expenses directly through the app and remind you when vaccinations or preventive care are due.

Automating Pet Care with Smart Home Integration

Smart home systems can also be integrated into your pet's care routine to automate their daily activities. Useful devices include the following:

- **Smart pet doors:** Pet doors like SureFlap Microchip Pet Door automatically open when they sense your pet's microchip, allowing them to come and go freely. This is perfect for pets that need access to a garden or yard and saves you from having to manually open the door every time. These doors can be set to open at specific times.

- **Smart feeders and camera integration:** Many smart pet devices can be linked to voice assistants. You can use voice commands to feed your pet, check in via a camera, or schedule feeding and playtime sessions.

Whether it's caring for pets, maintaining important relationships, or cutting down on manual meal preparation, automation can save you a ton of effort. Combine these with all the other methods we've covered, and you could find yourself with quite a bit more free time.

But that's just the start. In many areas, emerging AI technology is starting to transform the way we live, as we'll discuss next.

Key Takeaways

- **Automate regular shopping purchases**: Use subscription services like Amazon's Subscribe & Save to automatically reorder household essentials, pet supplies, and health products.

- **Integrate shopping and meal planning**: Platforms like Eat This Much offer end-to-end automation, syncing your meal plans with grocery lists and enabling direct ordering for delivery or in-store shopping, saving you both time and effort.

- **Automate relationship maintenance**: Schedule recurring reminders for birthdays, anniversaries, and regular check-ins with loved ones, using calendar apps or digital tools like Todoist to ensure meaningful connections are maintained even during busy periods.

- **Automate pet care**: Use smart devices like automatic feeders, pet exercise toys, and health monitoring collars to streamline pet care routines, from feeding and exercise to grooming and health tracking.

CHAPTER 11

AI'S ROLE IN AUTOMATING YOUR LIFE

You don't have to be an AI expert,
but you must be an AI explorer.
—Andrew Ng

Artificial intelligence (AI) has already begun to play a significant role in automating many aspects of our daily lives. As AI technology continues to evolve, it offers the potential to streamline tasks, improve efficiency, and help you make more informed decisions across various areas—from managing your finances to maintaining your health. AI's integration into automation makes life easier by handling repetitive tasks, allowing you to focus on what truly matters.

We've already covered examples of AI in previous sections. For instance, in the section about automating your home and environment, we discussed how smart thermostats can learn your habits, which uses AI. Another example is in the section about automating your finances—we discussed using AI to assist you with investing. In this chapter, we'll cover the reason it's important to learn about AI's impact on your day-to-day activities, plus additional ways AI can help automate your life.

AI Will Impact Your Daily Life

As AI becomes more integrated into our daily routines, understanding its capabilities and potential impact is no longer just a luxury for tech enthusiasts—it's a necessity for everyone. Learning how AI works and how it can automate various tasks empowers you to take full advantage of these technologies. It also helps you stay ahead of the curve in a world that's increasingly driven by AI.

First and foremost, understanding AI allows you to automate tasks that traditionally required manual effort. AI tools can take over repetitive, time-consuming tasks, freeing up your mental bandwidth for more important activities. As AI technology advances, these systems will become only more intuitive and offer increasingly efficient solutions that simplify your life.

But it's not just about convenience—it's also about being informed and in control. As AI continues to grow in influence, it's essential to understand how it can affect privacy, security, and decision-making. Knowing how AI systems function can help you recognize when those using the systems do it ethically and when they might be overstepping boundaries, whether that's through data collection, privacy concerns, or bias in decision-making.

Additionally, understanding AI's applications can help you make smarter and more informed choices. The knowledge you gain today can position you for success tomorrow in your financial, personal, and professional life. By recognizing how AI is reshaping environments, understanding its potential applications, and learning how to use AI-driven tools in your daily tasks, you can stay ahead in an increasingly competitive landscape.

Ultimately, learning about AI and its role in automating everyday tasks is more than just a tech trend—it's about reclaiming your time, optimizing your life, and empowering yourself to thrive in a world where technology is constantly evolving. That supports what this book is all about.

Ways AI Can Automate Your Life beyond the Basics

While AI is already making significant strides in simplifying tasks like budgeting or adjusting home temperatures, its potential reaches far beyond these common applications. In fact, AI can enhance disparate aspects of your life, from improving health and fitness to providing personalized recommendations and even offering assistance with personal care. As you look to streamline your daily routine and create more time for what matters, here are some advanced ways AI can help you automate and optimize your life.

Enhancing Your Smart Home with AI

AI can do much more than just control the temperature in your home. Integrating AI into your home automation system can create a truly intelligent environment that learns your needs, adapts to your lifestyle, and anticipates your desires.

For example, AI-powered home security systems can recognize faces; differentiate between people, pets, and objects; and alert you to unusual activity. These systems use machine-learning to reduce false alarms and become more accurate over time. Some systems can even integrate with doorbell cameras, enabling real-time facial recognition and personalized notifications and thus adding an extra layer of security and convenience.

Also, AI can enhance the management of other home devices. Smart kitchen assistants use AI to recommend recipes based on your dietary preferences or what's in your fridge. These devices can track inventory, place grocery orders automatically, or remind you when food is about to expire. When combined with your smart home assistant, these systems can become more than just a hub for controlling devices—they turn your home into a proactive environment that adapts to your needs.

Health and Fitness Assistants: Personalized and Real-Time Support

When it comes to health and fitness, AI has made it possible to automate monitoring, tracking, and even improvement recommendations without the need for constant input from you. Wearables like the Apple Watch, Fitbit, and Oura Ring are already using AI to monitor physical activity, sleep, heart rate, and blood oxygen levels. These devices don't just passively collect data; they actively analyze your patterns and provide personalized feedback. For instance, they can suggest adjustments to your fitness routine, detect early signs of overtraining, and remind you to get up and move when you've been sitting for too long.

AI-enhanced fitness apps, such as iFIT or JEFIT, go a step further by using machine-learning to create workout plans tailored to your goals. These apps not only generate personalized fitness routines but also adjust your training schedule, based on your progress, readiness, and performance. So, instead of manually tracking workouts or wondering what exercises to do next, you can use AI to tailor a personalized fitness journey and automatically update it to meet your needs.

In the realm of health diagnostics, AI is also making impacts. Applications like SkinVision use AI to analyze photos of your skin and flag potential concerns, like moles that could be early signs of skin cancer. Similarly, AI-powered health apps like Babylon Health offer virtual health consultations by using machine-learning algorithms to evaluate symptoms and provide medical advice. These technologies make healthcare more accessible and give you the tools to monitor your health and take preventive actions before issues escalate.

Personal Care AI Bots: Grooming and Self-Care Assistance

AI isn't limited to just fitness and health tracking; it can also play a role in personal care. Personal care AI bots enhance beauty routines by offering personalized skincare and makeup recommendations. These services use the camera in your phone and AI to analyze your skin's condition, taking into account factors like hydration levels, age, and environmental influences (e.g., sunlight and air pollution) to create a tailored skincare routine.

Also, AI-driven virtual assistants are now helping with grooming. Devices like Philips Smart Shaver use AI to analyze your skin and adapt their shaving technique to ensure a smoother, more comfortable shave. These tools make self-care simpler, more efficient, and highly personalized.

Additionally, AI-powered hair care systems, such as RevAir, personalize the drying process to match your hair type and texture, and give you healthier and shinier hair. These devices help take the hassle out of hair care, giving you more time to focus on other areas of your life.

Robo-Advisors: Smarter Investing with AI

Managing investments can be complex and time-consuming. However, robo-advisors powered by AI, such as Betterment, Wealthfront, and Ellevest, can automate much of the process. These platforms analyze your financial goals, risk tolerance, and market trends to make informed investment decisions on your behalf. Unlike traditional advisors, robo-advisors use algorithms to provide low-cost, accessible investment strategies that can be managed with minimal human intervention.

Robo-advisors can continuously monitor your portfolio; adjust your investments, based on market changes; and automatically rebalance your portfolio. This eliminates the need for regular check-ins or manual adjustments, saving you time. Over time, these tools learn from market patterns and user data, optimizing your portfolio management for both efficiency and long-term growth. A personalized investment manager who does similar work would be much more expensive.

Recommendation Engines: Shopping, Streaming, and Beyond

AI-powered recommendation engines have become a crucial part of how we shop and consume media. AI is behind the personalized recommendations you see from services like Netflix, YouTube, and Spotify. These engines analyze your preferences, purchase history, viewing habits, as well as what other users with similar tastes are enjoying, and then suggest products, movies, or songs that you're likely to enjoy.

For shopping, platforms like Amazon, Walmart, and eBay use AI to recommend products, based on your past purchases and browsing behavior. AI can also notify you about sales, upcoming trends, or new products that match your preferences. Alexa lets me know when items in my Amazon wish lists are on sale. It also asks me whether I need to purchase items I regularly consume if it thinks I may be running out, based on my past purchase frequency. Instead of spending time sifting through hundreds of options, you can use AI to curate a personalized shopping experience, making it easier to find what you want.

Automating Household Chores: Robots and AI Assistants

Household chores are one of the most time-consuming aspects of daily life, but AI-powered robots are changing that. I mentioned robot vacuums earlier in this book, but I didn't explain their connection to AI. Many of them use AI to map out your home, navigate obstacles, and clean more efficiently over time. These robots learn the layout of your home, identify high-traffic areas, and adapt to different floor types.

More advanced systems, such as robotic lawnmowers and pool cleaners, are also using AI to handle outdoor chores. These devices can autonomously cut your grass or clean your pool, saving you hours of work each week. And, with integration into your smart home, you can set schedules, track progress, and even control these devices remotely from your smartphone.

AI in Education and Learning

Learning can be personalized and automated using AI, which helps students and professionals alike make the most of their time. AI-powered educational platforms like Khan Academy and Coursera use algorithms to recommend courses based on your learning history, goals, and preferences. AI can also offer personalized tutoring, identifying areas where you need improvement and adapting the content to suit your needs.

In professional settings, AI can optimize your career growth by recommending skills or courses aligned with your goals. AI can help advance your knowledge and stay competitive in your field.

What I Do

I use AI in many facets of my personal and professional life:

- I use a Polar fitness watch to track my workouts and overall activity. It uses AI to recommend workouts to me, as well as various information about my heart rate, sleep, and breathing to recommend how hard I should work out on a given day.

- I use ChatGPT for financial planning. I also use it to discuss topics for books and articles I write, and to help fine-tune the language of my writings. I've even integrated it with my smart home controller to have it answer questions and control my smart home.

- I own a couple of robot vacuums that use AI to map and clean my floors while avoiding objects and not falling off the stairs.

- My ecobee smart thermostat learns over time when it needs to turn on to get to temperature by a certain time.

- I use AI social media tools to help me generate social media posts and ideas for my business.

- Like you, I get show and movie recommendations on Netflix, YouTube TV, and other streaming services, which are generated by AI. AI also recommends songs and playlists on my streaming music service.

It's clear AI has given existing software tools a turbo boost, helping you improve your most important skills while replacing tedious chores altogether. In the final chapter, we'll look at how you can put together all of these learnings in this book to truly transform the way you live.

Key Takeaways

- **Understand AI's impact:** Knowledge of AI's capabilities helps you make informed decisions, protect your privacy, and stay ahead in an increasingly AI-driven world.

- **AI in health and fitness:** AI-powered wearables and apps personalize fitness routines, monitor health metrics (e.g., heart rate, sleep, blood oxygen), and provide real-time feedback, enabling users to optimize workouts and track overall wellness without constant input.

- **Robo-advisors and investment automation:** Robo-advisors use AI to manage investments, automatically adjust portfolios, and rebalance assets, providing low-cost, automated, and efficient financial management for users.

- **AI in education and learning:** AI personalizes the learning experience by recommending courses, identifying areas of improvement, and tailoring content to suit your needs.

CHAPTER 12

PUTTING IT ALL TOGETHER: YOUR NEW LIFE BY DESIGN

In the end we retain from our studies
only that which we practically apply.
—*Johann Wolfgang Von Goethe*

We've covered a lot of ground in this book, and it is okay, even normal, if you feel overwhelmed. We've looked at a large number of ideas and tools that you can apply to pretty much every area of life. But where on earth do you begin, and how do you prioritize one area over another?

The point of automation isn't to fry your brain with the myriad of possibilities; it is to make things simpler, precisely so you don't have to think about them! For the best results, do the following: (1) have routines, (2) occasionally review and optimize your routines, and (3) prioritize making improvements to the things that are really going to help.

Start and End Each Day with Purpose and Routine

Your day begins and ends with intention. Setting a clear purpose at the start of your day and reflecting at the end are simple but powerful practices. Use automation to kickstart this routine and keep you on track. For example, schedule a daily review of your to-do list every morning.

Also, keep in mind the tasks that need doing each day and that can be automated. Maybe it's vacuuming a high-traffic area of home, using a robot vacuum set to clean on schedule. Perhaps it's something even simpler like setting up "Do Not Disturb" times on your phone for reflection, family time, and work. Think about what tasks can be automated each day and set forth toward automating them.

At the end of your day, an automated system can help you reflect on your progress. Set a reminder to review your accomplishments and identify areas where you could improve. This reflection process, paired with automated tracking of your activities, helps you move toward your goals, no matter how hectic life gets. Part of your reflection process should include explicitly listing and reviewing your top accomplishments. These will help motivate you to keep going.

Start and End Each Month with Purpose and Routine

Monthly reviews offer a broader view of your progress and help you adjust your course when needed. Use automation to set aside time at the beginning and end of each month to assess your goals. If you've already set up an automated financial tracking tool like Monarch Money or Simplifi, you can instantly see whether you've kept to your budget in the last month and are closer to your savings goals.

Use your calendar to set recurring monthly appointments for personal reflection. Set up regular meetings with loved ones to stay in touch. Perhaps automate reminders to update your goals, track your personal growth, and organize your priorities for the coming month. Monthly reviews help you recalibrate, allowing you to shift gears if necessary to make sure you're spending your time and energy on the right things.

Start and End Your Year with Purpose

At the end of the year, it's important to reflect on what you've accomplished and plan for the year ahead. Use AI tools and automation to pull together data from various aspects of your life—financial, fitness, and personal growth—and create a comprehensive yearly review. Apps like Google Calendar and Notion can help you visualize your year, track milestones, and make data-driven decisions about your goals for the next year.

Just as with your monthly reviews, schedule time at the start of each year to review your bigger goals. Set up reminders in advance so that, when the new year arrives, you're ready to hit the ground running, fully prepared with a clear vision for the year ahead.

Keep Track of Automated Tasks and Reminders in One Place

I recommend keeping documentation of your life automations in one place. A centralized calendar or to-do list tool (Google Calendar, Todoist, Notion, etc.) can sync all your automated reminders, tasks, and reflections. This creates a single source of truth for everything you need to stay on top of in your personal and professional life. These tools allow you to view your schedule at a glance, track your goals, and adjust as needed—all in one place.

Automate What You Don't Care to Spend Much Time On

Throughout this book, you've learned about numerous areas where automation can save you time—whether it's in your smart home, fitness routine, health monitoring, or personal finances. The key to success is to focus on automating the tasks you don't need to dedicate much time or mental energy to. This gives you the freedom to focus on the areas that matter most to you—your relationships, passions, and long-term goals.

Look through the "What I Do" and "In Real Life" sections of this book for inspiration. Be sure to review the tools and resources in the glossary. Then, choose the tools that best fit your needs and you'll free up hours each week that can be used to focus on things that align with your values.

Finding the Right Balance

While automation is a powerful tool, the key to a successful life by design is balance. It's easy to get carried away with automating everything, but automation should enhance your life, not take over. Find the balance between using technology to save time and maintaining the human touch in areas that matter most—your relationships, creative pursuits, and personal growth.

Start with automating low-priority tasks and gradually expand as you become more comfortable with the systems you set up. Evaluate how much time you're saving and whether that time is being used effectively to invest in the things that bring you joy and fulfillment. Remember, automation should be a tool that works for you—not a replacement for living intentionally.

Your New Life by Design

Putting it all together means creating a system that works for you—one that balances efficiency with intentionality. By using the tools, systems, and practices outlined throughout this book, you'll be able to automate the tasks that drain your time and energy, giving you more space to focus on the things you truly care about. Over time, you'll build a life that is not only easier but also more fulfilling. With AI and automation on your side, you'll have the time and freedom to live your best life—one that's centered around what matters most to you.

NEXT STEPS

First of all, thank you for purchasing *Life by Design*. I know that you could have picked any number of books to read, but you've picked this book, and for that I am extremely appreciative.

I hope that it has inspired you and helped you spend time on the things in your life that are most important to you. If so, it would be really nice if you could share this book with your friends and family by **posting about it on Twitter, Instagram, Facebook, and Pinterest.**

If you enjoyed this book and found some benefit from reading it, I'd love to hear from you. I hope that you can take the time to **post a review on Amazon.** Your feedback and support will help me greatly improve my writing craft for future projects.

I want you, the reader, to know that your review is very important and very much appreciated.

I wish you all the best in your journey to focus on what matters most in your life!

Subscribe to HomeTechHacker.com

Be sure to check out HomeTechHacker.com for free in-depth articles about automation, home networks, cybersecurity, and more. Also, subscribe to my newsletter to get the latest updates about automating your life and other home technology topics.

Consider Marlon Buchanan's Other Books

- *The Personal Cybersecurity Manual: How Anyone Can Protect Themselves from Fraud, Identity Theft, and Other Cybercrimes*

- *The Home Network Manual: The Complete Guide to Setting Up, Upgrading, and Securing Your Home Network*

- *The Smart Home Manual: How to Automate Your Home to Keep Your Family Entertained, Comfortable, and Safe*

- *Home Wi-Fi Tuneup: Practical Steps You Can Take to Speed Up, Stabilize, and Secure Your Home Wi-Fi*

All books are available in e-book and paperback formats. *The Personal Cybersecurity Manual*, *The Home Network Manual*, and *The Smart Home Manual* are also available as audiobooks. You can learn more about these books, including where to buy them, at MarlonBuchanan.com.

TOOLS AND RESOURCES

HomeTechHacker Resources

- **HomeTechHacker Blog** (https://hometechhacker.com/): My personal blog where you can find many helpful articles about automation and ways to improve your home network and personal cybersecurity practices.

- **HomeTechHacker Shop** (https://hometechhacker.com/shop/): Here, I maintain an up-to-date list of recommended home network and smart home devices.

- **HomeTechHacker Academy** (https://academy.hometechhacker.com): Practical, easy-to-follow self-paced online courses on smart home technology, home networking, artificial intelligence, and personal cybersecurity.

Task Management Resources

- **Trello** (https://trello.com/): A visual project management tool that organizes tasks into boards and cards for easy tracking and collaboration.

- **Todoist** (https://todoist.com/): A task manager that helps you create, organize, and automate your to-do lists for better productivity.

- **Notion** (https://www.notion.com/): All-in-one workspace that blends everyday work apps into one, with customizable building blocks for notes, tasks, wikis, and databases.

- **Google Keep** (https://keep.google.com/): A note-taking service for creating and organizing notes, lists, photos, and audio memos, accessible across various devices.

Financial Management Resources

- **YNAB** (https://www.youneedabudget.com/): App for taking control of your finances with budgeting tools and automated expense tracking.

- **Monarch Money** (https://www.monarchmoney.com/): Personal finance platform that helps you track spending, create budgets, monitor investments, and collaborate with partners in one place.

Shopping Resources

- **Giftagram** (https://www.giftagram.com/): Gifting platform offering a curated selection of products from various brands, allowing you to send gifts without needing the recipient's address.

- **Amazon Subscribe & Save** (https://www.amazon.com/Subscribe-Save/b?ie=UTF8&node=5856181011): Schedule recurring deliveries of eligible products and receive discounts on those items.

- **Instacart** (https://www.instacart.com): Order groceries online from local stores and have them delivered to your door, with the option to schedule deliveries and shop from a variety of retailers.

Grocery and Meal Planning Resources

- **Vitamin Shoppe** (https://www.vitaminshoppe.com): A retailer offering a wide range of vitamins, supplements, and health products, with options for online shopping, in-store pickup, and personalized recommendations to support various wellness goals.

- **Amazon Fresh** (https://www.amazon.com/fresh): Amazon's grocery delivery service offers a wide selection of fresh produce, pantry staples, and household items, available for delivery or pickup.

- **AnyList** (https://www.anylist.com): Grocery shopping and recipe organization app that allows you to create shopping lists, save recipes, and share lists with others for easier meal planning and grocery shopping.

- **Out of Milk** (https://www.outofmilk.com): Shopping list and pantry management app for keeping track of items you already have or need to get, with the ability to organize shopping lists by category for extra efficiency.

- **Flipp** (https://www.flipp.com): An app that aggregates digital flyers and coupons from major retailers, so you can browse and plan your shopping to find the best deals on groceries and household items.

- **Plan to Eat** (https://www.plantoeat.com): Create customized meal plans, organize recipes, and generate shopping lists based on your chosen meals, streamlining the planning and shopping process.

- **Blue Apron** (https://www.blueapron.com): A meal kit delivery service that provides pre-portioned ingredients and recipes for home cooking, helping you create fresh, home-cooked meals with ease.

- **Home Chef** (https://www.homechef.com): A meal kit delivery service that offers a variety of easy-to-cook meal options, including customizable ingredients and quick cooking times, delivered right to your door.

- **Hello Fresh** (http://www.hellofresh.com/): A meal delivery service supplying pre-measured ingredients and step-by-step recipes to your doorstep.

- **Mealime** (https://www.mealime.com/): Helps you plan meals by generating personalized recipes and automated grocery lists based on your preferences.

- **Paprika** (https://www.paprikaapp.com): A recipe management app that lets you save and organize recipes, create meal plans, generate shopping lists, and sync data across multiple devices for a seamless cooking experience.

- **Cookpad** (https://www.cookpad.com): A recipe-sharing platform that allows you to discover, share, and save homemade recipes, offering a community-driven approach to cooking and meal inspiration.

- **Eat This Much** (https://www.eatthismuch.com): A meal-planning and calorie-tracking app that generates weekly meal plans and shopping lists based on your nutritional needs, preferences, and goals, taking the guesswork out of meal prep.

Smart Home and Home Automation Resources

- **Home Assistant** (https://www.home-assistant.io/): Open-source platform for automating smart home devices. It provides a wealth of tutorials, guides, and resources on how to automate everything from lights to security systems.

- **Automate the Boring Stuff with Python** (https://automatetheboringstuff.com/): This website and accompanying book by Al Sweigart teaches you how to use Python and other tools to automate everyday tasks like renaming files, sending emails, scraping websites, and more—ideal for those looking to dive into the world of coding for automation.

Charity and Giving Resources

- **JustGiving** (https://www.justgiving.com/): An online fundraising platform that helps individuals, charities, and organizations raise money for various causes. It allows you to set up fundraising pages, collect donations, and track progress, with a focus on charity-driven campaigns.

- **GoFundMe Charity** (https://www.gofundme.com/c/start/charity-fundraising): A platform that enables individuals and organizations to create fundraising campaigns for charitable causes. You can easily raise money for personal, community, or nonprofit projects, with no platform fees for charity fundraising.

Fitness and Health Resources

- **MyFitnessPal** (https://www.myfitnesspal.com/): A popular mobile app that helps you track your food intake, exercise, and nutrition goals. It provides a large database of foods, offers personalized recommendations, and integrates with other fitness apps to support weight loss and health management.

- **Lose It!** (https://www.loseit.com/): A user-friendly app that helps you track your calories and exercise. It features a food database, barcode scanning, and personalized weight loss plans to make it easier to achieve your health and fitness goals.

- **Calm** (https://www.calm.com/): A meditation and relaxation app offering guided sessions, sleep stories, breathing exercises, and calming music to reduce stress, improve focus, and promote better sleep.

- **Headspace** (https://www.headspace.com/): A popular meditation app that provides guided mindfulness and meditation practices aimed at reducing stress, improving mental clarity, and fostering emotional well-being.

- **Insight Timer** (https://insighttimer.com/): A free meditation app with a vast library of guided sessions, music, and soundscapes. It also includes a customizable meditation timer and a community of users for shared mindfulness practices.

- **iFIT** (https://www.ifit.com): A fitness subscription service that offers personalized workout programs for various types of fitness equipment, including treadmills, ellipticals, and bikes, along with virtual coaching and access to global workout environments.

- **JEFIT** (https://www.jefit.com): A workout tracking and fitness app that lets you log workouts, create customized fitness plans, track your progress, and access a library of exercises for strength training and bodybuilding.

Home Maintenance Resources

- **HomeZada** (https://www.homezada.com/): A digital home management platform that helps homeowners organize and track their home maintenance, inventory, and improvement projects, while also managing budgets and tasks.

- **Househappy** (https://www.househappy.com/): An online platform that connects homeowners with local service providers for home improvement and maintenance needs, offering reviews and recommendations to simplify the process of finding reliable professionals.

- **FilterEasy** (https://www.filtereasy.com): A subscription service that delivers air filters to your door on a regular schedule, helping you maintain clean air and improve HVAC system efficiency by automatically replacing filters when needed.

Travel Resources

- **TripIt** (https://www.tripit.com): Travel organization app that consolidates travel plans into a single itinerary, automatically importing travel details from emails and providing real-time alerts and reminders to keep you on track during your trips.

- **AwardWallet** (https://www.awardwallet.com): Loyalty program tracker for managing and monitoring your frequent flyer miles, hotel points, and other rewards. It provides easy access to balances and expiration dates for multiple programs in one place.

- **ParkWhiz** (https://www.parkwhiz.com): A parking reservation service that lets you find and book parking spots in advance at discounted rates, saving time and avoiding a search for parking in busy areas.

- **SpotHero** (https://www.spothero.com): Find, reserve, and pay for parking in advance at thousands of locations with this app. Offers discounted rates and convenient parking options in cities across the U.S.

- **PackPoint** (https://www.packpnt.com): App that helps you organize and customize your travel packing, based on factors like destination, weather, and trip duration, ensuring you don't forget any essentials for your journey.

- **FlightAware** (https://www.flightaware.com): Flight-tracking website and app that provides real-time information on flight status, including arrival and departure times, delays, cancellations, and live flight-tracking across the globe.

- **Google Flights** (https://www.google.com/flights): A travel search engine for finding, comparing, and booking flights, with tools to track prices, explore destinations, and set up price alerts for future trips.

Job Search Resources

- **Dux-Soup** (https://www.dux-soup.com): A LinkedIn tool for managing lead generation and outreach by automating profile visits, connection requests, and messaging.

- **Jobscan** (https://www.jobscan.co): A tool that optimizes your résumé by comparing it to job descriptions and helps you receive actionable recommendations to improve your chances of passing through applicant tracking systems (ATS).

- **Codility** (https://www.codility.com): A platform offering coding assessments and challenges for recruiters and hiring managers, enabling them to evaluate candidates' technical skills through real-world coding tests.

- **HackerRank** (https://www.hackerrank.com): A platform that allows you to improve your coding skills through challenges and competitions. The platform also offers companies a way to assess potential hires' abilities through technical tests.

- **LeetCode** (https://www.leetcode.com): A platform that allows you to try coding challenges and algorithm practice problems to improve your problem-solving skills in preparation for technical interviews.

- **Vervoe** (https://www.vervoe.com): A skills assessment platform that allows employers to create custom tests and simulate real job tasks. This helps them hire the best candidates, based on demonstrated capabilities rather than résumés alone.

- **Big Interview** (https://www.biginterview.com): Improve your job interview skills and build your confidence through practice video sessions, coaching, expert advice, and simulated scenarios.

Pet Care Resources

- **Chewy** (https://www.chewy.com): Retailer specializing in pet food, supplies, and medications. It offers convenient home delivery, personalized pet product recommendations, and auto-ship options for recurring orders.

- **Trupanion** (https://www.trupanion.com): Pet insurer offering comprehensive coverage for unexpected veterinary costs, with a focus on providing direct payments to vets at time of treatment for faster claims processing.

- **Fetch** (https://www.fetchpet.com/): Pet insurance company covering a wide range of treatments and conditions, with customizable plans to help you manage the cost of veterinary care for your dogs and cats.

Other Resources

- **AI For Everyone Course** (https://www.coursera.org/learn/ai-for-everyone): This free course by Andrew Ng provides a non-technical introduction to artificial intelligence and its applications in everyday life.

- **1Password** (https://www.1password.com): Password manager that securely stores and manages passwords, credit card information, and other sensitive data, with features like password generation, encryption, and cross-device syncing to ensure safe and easy access.

- **Feedly** (https://www.feedly.com): Content aggregation tool for following your favorite blogs, news sites, and online publications in one place, thereby streamlining your content consumption and keeping you updated on topics of interest.

- **Zapier** (https://www.zapier.com): An automation tool that connects over five thousand apps and automates workflows by creating "Zaps"—triggers and actions that save time by performing repetitive tasks without manual intervention.

- **Birchbox** (https://www.birchbox.com): A subscription service that delivers personalized beauty and grooming samples each month, helping you discover new skincare, haircare, and makeup products tailored to your preferences.

- **IPSY** (https://www.ipsy.com): A beauty subscription service that sends monthly bags filled with personalized samples of skincare, makeup, and tools, based on your preferences and beauty profile.

INSPIRATIONAL QUOTES

I'm a big fan of inspirational, profound, and informative quotes. Here are some quotes that I hope will inspire you on your journey to automate your life and to focus on what's really important.

Motivation Quotes

"The secret of getting ahead is getting started. The secret to getting started is breaking your complex, overwhelming tasks into small manageable tasks and then starting on the first one." —Mark Twain

"Motivation is what gets you started. Habit is what keeps you going." —Jim Rohn

Time Management Quotes

"We're all so busy we don't make time to enjoy our lives, good company and good food. There are only the pursued, the pursuing, the busy and the tired." —John Torode

"I wanted to figure out why I was so busy, but I couldn't find the time to do it." —Todd Stocker

"The root of productivity is in personal priorities. Know what matters to you and why." —Melissa Steginus

"The first step to regain control of time is to decide what activities are most important so that we can plan to give them

the proper priority during a day or a week or a month."
—Charles E. Hummel

"Time management is an oxymoron. Time is beyond our control, and the clock keeps ticking regardless of how we lead our lives. Priority management is the answer to maximizing the time we have." —John C. Maxwell

"You'll never change your life until you change something you do daily. The secret of your success is found in your daily routine." —John C. Maxwell

Financial Management Quotes

"A big part of financial freedom is having your heart and mind free from worry about the what-ifs of life." —Suze Orman

"Money, like emotions, is something you must control to keep your life on the right track." —Natasha Munson

"An investment in knowledge always pays the best interest." —Benjamin Franklin

Fitness and Health Quotes

"Take care of your body. It's the only place you have to live." —Jim Rohn

"Know that what seems really hard today, with good practice and consistency, will be a piece of cake later." —Massy Arias

"The future of fitness is personal, precise, and automated." —Rishi Mandal

"Personalize your progress, automate your performance." —Unknown

"The secret to a healthier life is not in the intensity of effort but in the consistency of automated habits." —Unknown

Automation Quotes

"There's a lot of automation that can happen that isn't a replacement of humans, but of mind-numbing behavior." —Stewart Butterfield

"Automation isn't about creating a lazy lifestyle; it's about creating a more efficient and enjoyable one." —Unknown

"Automation is driving the decline of banal and repetitive tasks." —Amber Rudd

"You're either the one that creates the automation or you're getting automated." —Tom Preston-Werner

"Automation is not a thing of the future, but a thing of the present." —Brian Tracy

"Automation is the new electricity. It's transformative, and it's going to change everything." —Ken Goldberg

"Automation is not a one-time event, but a continuous process of improvement and innovation." —Anand Deshpande

"Automation is not about doing more with less, but about doing more with the resources we have." —Antonio Grasso

Cybersecurity Quotes

"Choosing a hard-to-guess, but easy-to-remember password is important!" —Kevin Mitnick

"A password manager is your digital fortress." —Unknown

"Your password is your digital identity. Don't let it be a punchline." —Unknown

Artificial Intelligence Quotes

"Artificial intelligence is the future, and the future is here."
—Fei-Fei Li

"AI is not just a tool for automation; it's an enabler for augmentation." —Satya Nadella

"AI is going to be a part of our lives and will help us solve many problems, but only if we let it." —Demis Hassabis

"AI won't replace humans, but those who use AI will replace those who don't." —Garry Kasparov

Other Inspirational Quotes

"The future belongs to those who see possibilities before they become obvious." —John Sculley

"Knowledge is not power. Action is power." —Tony Robbins

"There is no greater harm than that of time wasted."
—Michelangelo

"Every accomplishment starts with a decision to try." —
Gail Devers

"The great aim of education is not knowledge but action."
—Herbert Spencer

"Change is the end result of all true learning." —Leo
Buscaglia

GLOSSARY

2FA (Two-Factor Authentication): A security method that requires two forms of identification, such as a password and a verification code, to authenticate access to an account.

AI (Artificial Intelligence): Technology that simulates human intelligence, enabling systems to perform tasks such as learning, reasoning, and problem-solving.

Applicant Tracking System (ATS): Software that automates the hiring process for employers by managing job postings, résumés, and applications. It streamlines recruitment by sorting, filtering, and ranking candidates, based on specific criteria, improving efficiency and reducing manual work.

Attention Economy: A system in which human attention is treated as a scarce resource, with businesses and content creators competing to capture and retain it, often through digital platforms, advertisements, and engaging content.

Authenticator App: A mobile application that generates time-sensitive authentication codes used for securing online accounts through 2FA.

Automatic Bill Payment: The set up of recurring payments for bills that are automatically processed by a bank or service provider.

Automation: The use of technology to perform tasks without human intervention.

Biometric Authentication: A security process that uses an individual's unique physical characteristics, such as fingerprints, facial recognition, or iris scans, to verify their identity and grant access to devices or systems.

CLEAR: A trusted traveler program that uses biometric identification (such as fingerprints and retina scans) to expedite airport security screening, offering a faster, contactless experience for travelers.

Cybersecurity: The practice of protecting computer networks, devices, and data from unauthorized or criminal use. This book focuses on *personal* cybersecurity, which is the practices individuals and families should take to protect themselves from cybercriminals.

Exchange-Traded Fund (ETF): A type of investment fund that holds a diversified portfolio of assets, such as stocks or bonds, and trades on stock exchanges like individual stocks. ETFs offer investors a low-cost, flexible way to invest in a broad market index or specific sector.

FOMO (Fear of Missing Out): The anxiety or fear that one is missing out on something exciting or valuable, often triggered by seeing others' activities or experiences, especially on social media.

Geofencing: A location-based service that uses GPS or RFID technology to create virtual boundaries around a physical area. When a device enters or exits this designated area, it triggers a predefined action, such as sending a notification, activating an app, or controlling smart devices.

Global Entry: A U.S. Customs and Border Protection program that allows expedited clearance for pre-approved, low-risk travelers entering the United States, using automated kiosks at select airports.

Hardware Security Keys: Physical devices used for two-factor authentication (2FA) that generate or store authentication credentials, offering a secure and tamper-resistant method for verifying a user's identity when accessing online services.

Heart Rate Variability (HRV): The measure of the variation in time between each heartbeat. It reflects the health and flexibility of the autonomic nervous system, with higher HRV generally indicating better cardiovascular fitness and stress resilience. HRV is often used to assess recovery and overall well-being.

Home Assistant: An open-source platform for automating smart home devices, providing resources to control and monitor various home systems.

HVAC (Heating, Ventilation, and Air Conditioning): A system used to control the temperature, humidity, and air quality in buildings, ensuring comfort and health for occupants.

OBD II Port: A standardized diagnostic interface found in most vehicles manufactured after 1996, allowing mechanics and technicians to connect a diagnostic tool to the vehicle's onboard computer system for reading trouble codes, monitoring vehicle performance, and diagnosing issues.

Password Manager: A software tool that stores and manages your passwords in an encrypted database, often with the ability to automatically fill in passwords on websites.

Recommendation Engine: A system that uses algorithms to suggest products, services, or content to users, based on their preferences, behaviors, or similar user patterns, commonly used in platforms like Netflix or Amazon.

Robo-Advisor: An AI-driven platform that provides automated financial advice based on algorithms, typically used for investment management.

RSS Feeds: A format for delivering regularly updated content from websites, blogs, and news sources. **RSS** (Really Simple Syndication) allows users to subscribe to feeds and receive automatic updates in a feed reader, making it easier to stay up to date with new content without visiting each website individually.

SMART Goals: A framework for setting clear and achievable objectives, where each goal is characterized as follows:

- Specific: Clearly defines the goal and its purpose.
- Measurable: Includes criteria to track progress and determine success.
- Achievable: Is realistic and attainable, given available resources.
- Relevant: Is aligned with broader personal or professional objectives.
- Time-bound: Has a defined deadline for completion.

Smart Home: A home equipped with internet-connected devices that can be controlled remotely for enhanced convenience, security, and energy efficiency.

TSA PreCheck: A trusted traveler program in the U.S. for expedited airport security screening, reducing wait times by not requiring passengers to remove shoes, belts, or laptops.

Zelle: A peer-to-peer payment service that allows users to send money directly from their bank accounts, using an email address or mobile number.

ABOUT THE AUTHOR

Marlon Buchanan, bestselling author and IT Director, holds a bachelor's degree in Computer Science and Engineering from MIT and master's degrees in Business Administration and Software Engineering from Seattle University. With over 25 years of experience in IT, software development, and smart home automation, he helps busy professionals harness technology to create more time, reduce stress, and design lives they love.

He has been automating things around his house since he was a kid. Now, his wife and kids get to enjoy the fruits of his automation and efficiency exploits. He is best known for his smart home, home networking, and cybersecurity articles on his blog HomeTechHacker.com, and his bestselling books *The Personal Cybersecurity Manual*, *The Home Network Manual*, and *The Smart Home Manual*.

Please sign up for his newsletter on his blog. You can also follow him on these social media channels (@ HomeTechHacker):

- X (Twitter): twitter.com/HomeTechHacker
- Instagram: instagram.com/HomeTechHacker/
- Pinterest: pinterest.com/HomeTechHacker
- Facebook: facebook.com/HomeTechHacker